Communications in Computer and Information Science   967

Commenced Publication in 2007
Founding and Former Series Editors:
Phoebe Chen, Alfredo Cuzzocrea, Xiaoyong Du, Orhun Kara, Ting Liu,
Krishna M. Sivalingam, Dominik Ślęzak, and Xiaokang Yang

Editorial Board

Anna Monreale · Carlos Alzate et al. (Eds.)

ECML PKDD 2018 Workshops

Workshops

DMLE 2018 and IoTStream 2018
Dublin, Ireland, September 10–14, 2018
Revised Selected Papers

Springer

Editors
Anna Monreale (iD)
KDDLab
University of Pisa
Pisa, Pisa, Italy

Carlos Alzate (iD)
IBM Research - Ireland
Dublin, Ireland

Workshop Editors *see next page*

ISSN 1865-0929   ISSN 1865-0937   (electronic)
Communications in Computer and Information Science
ISBN 978-3-030-14879-9   ISBN 978-3-030-14880-5   (eBook)
https://doi.org/10.1007/978-3-030-14880-5

Library of Congress Control Number: 2019933158

This Springer imprint is published by the registered company Springer Nature Switzerland AG
The registered company address is: Gewerbestrasse 11, 6330 Cham, Switzerland

Workshop Editors

Michael Kamp (iD)
University of Bonn
Bonn, Nordrhein-Westfalen, Germany

Daniel Paurat (iD)
Fraunhofer Institute for Intelligent Ana
Sankt Augustin, Nordrhein-Westfalen
Germany

Albert Bifet
ParisTech
LTCI, Telecom
Paris, France

Rita P. Ribeiro
Laboratory of Artificial Intelligence
and Decision Support
University of Porto
Porto, Portugal

Yamuna Krishnamurthy (iD)
Royal Holloway University of London
Egham, UK

Moamar Sayed-Mouchaweh
Départment Informatique et Automati
École des Mines de Douai
Reims, France

João Gama (iD)
Laboratory of Artificial Intelligence
and Decision Support
University of Porto
Porto, Portugal

Preface

The European Conference on Machine Learning and Principles and Practice of Knowledge Discovery in Databases (ECML PKDD) is the premier European machine learning and data mining conference and builds upon over 16 years of successful events and conferences held across Europe. ECML PKDD 2018 was held in Dublin, Ireland, during September 10–14, 2018. It was complemented by a workshop program, where each workshop was dedicated to specialized topics, cross-cutting issues, and upcoming trends. This year, 19 workshop proposals were submitted, and after a careful review process led by the workshop co-chairs, 17 workshops were accepted. The workshop program included the following workshops:

1. The Third Workshop on Mining Data for Financial Applications (MIDAS)
2. The Second International Workshop on Personal Analytics and Privacy (PAP)
3. New Frontiers in Mining Complex Patterns
4. Data Analytics for Renewable Energy Integration (DARE)
5. Interactive Adaptive Learning
6. The Second International Workshop on Knowledge Discovery from Mobility and Transportation Systems (KnowMe)
7. Learning with Imbalanced Domains: Theory and Applications
8. IoT Large-Scale Machine Learning from Data Streams
9. Artificial Intelligence in Security
10. Data Science for Human Capital Management
11. Advanced Analytics and Learning on Temporal Data
12. The Third Workshop on Data Science for Social Good (SoGood)
13. Urban Reasoning from Complex Challenges in Cities
14. Green Data Mining, International Workshop on Energy Efficient Data Mining and Knowledge Discovery
15. Decentralized Machine Learning on the Edge
16. Nemesis 2018: Recent Advances in Adversarial Machine Learning
17. Machine Learning and Data Mining for Sports Analytics (MLSA)

Each workshop had an independent Program Committee, which was in charge of selecting the papers. The success of the ECML PKDD 2018 workshops depended on the work of many individuals. We thank all workshop organizers and reviewers for the time and effort invested. We would also like to express our gratitude to the members of the Organizing Committee and the local staff who helped us. Sincere thanks are due to Springer for their help in publishing the proceedings.

This volume includes the selected papers of the Decentralized Machine Learning on the Edge and IoT Large-Scale Machine Learning from Data Streams workshops. The papers of the other workshops will be published in separate volumes. Lastly, we thank all participants and keynote speakers of the ECML PKDD 2018 workshops for their contributions that made the meeting really interesting.

December 2018
Carlos Alzate
Anna Monreale

Contents

Decentralized Machine Learning
on the Edge

Preface

Decentralized Machine Learning on the Edge

Many of today's parallel machine learning algorithms were developed for tightly coupled systems like computing clusters or clouds. However, the volumes of data generated from machine-to-machine interaction, by mobile phones or autonomous vehicles, surpass the amount of data that can be realistically centralized. Thus, traditional cloud computing approaches are rendered infeasible.

To scale parallel machine learning to such volumes of data, computation needs to be pushed towards the edge, that is, towards the data generating devices. By learning models directly on the data sources - which often have computational power of their own, for example, mobile phones, smart sensors, and tablets - network communication can be reduced by orders of magnitude. Moreover, it enables training a central model without centralizing privacy-sensitive data. The Decentralized Machine Learning at the Edge (DMLE'18) workshop aimed to foster discussion, discovery, and dissemination of novel ideas and approaches for decentralized machine learning.

The first international DMLE'18 workshop was held in Dublin, Ireland in conjunction with ECMLPKDD. The workshop included a keynote by Dr. Michael May (Siemens AG) followed by technical presentations and a poster session. The workshop was attended by around 30 people.

The accepted papers presented interesting novel aspects of decentralized machine learning, especially in the context of edge computing. We want to thank the authors for their valuable contributions, great presentations, and lively and fruitful discussions.

We would also like to thank the DMLE'18 program committee, whose members made the workshop possible with their rigorous and timely review process. Finally, we would like to thank ECMLKPDD for hosting the workshop and the workshop chairs, Anna Monreale and Carlos Alzate for their valuable support.

Organization

DMLE'18 Chairs

Michael Kamp   University of Bonn & Fraunhofer IAIS
Yamuna Krishnamurthy   Royal Holloway University of London
Daniel Paurat   Fraunhofer IAIS

Program Committee

Katharina Morik   TU Dortmund, Germany
Stefan Wrobel   Fraunhofer IAIS, Germany
Tamas Horvath   University of Bonn, Germany
Mario Boley   Max Planck Institute for Informatics,
     and Saarland University, Germany
Sandy Moens   University of Antwerp, Belgium
Janis Keuper   Fraunhofer ITWM, Germany
Dino Oglic   University of Nottingham, UK
Rafet Sifa   Fraunhofer IAIS, Germany

Sparsity in Deep Neural Networks - An Empirical Investigation with TensorQuant

Dominik Marek Loroch[1,2(✉)], Franz-Josef Pfreundt[1], Norbert Wehn[2], and Janis Keuper[1,3]

[1] Fraunhofer ITWM, Kaiserslautern, Germany
dominik.loroch@itwm.fhg.de
[2] TU Kaiserslautern, Kaiserslautern, Germany
[3] Fraunhofer Center Machine Learning, St. Augustin, Germany

Abstract. Deep learning is finding its way into the embedded world with applications such as autonomous driving, smart sensors and augmented reality. However, the computation of deep neural networks is demanding in energy, compute power and memory. Various approaches have been investigated to reduce the necessary resources, one of which is to leverage the sparsity occurring in deep neural networks due to the high levels of redundancy in the network parameters. It has been shown that sparsity can be promoted specifically and the achieved sparsity can be very high. But in many cases the methods are evaluated on rather small topologies. It is not clear if the results transfer onto deeper topologies.

In this paper, the *TensorQuant* toolbox has been extended to offer a platform to investigate sparsity, especially in deeper models. Several practical relevant topologies for varying classification problem sizes are investigated to show the differences in sparsity for activations, weights and gradients.

Keywords: Deep neural networks · Sparsity · Toolbox

1 Introduction

For the past few years, deep learning had a high impact on machine learning. Many diverse applications have emerged in virtually all fields of research and everyday life. Initially being a high-performance computing problem, deep learning is finding its way into the mobile and embedded world with applications such as autonomous driving, smart sensors and augmented reality, just to name a few. There is a huge potential behind deep learning in the embedded world, where more and more of the heavy workload is moved to the device, known as edge computing.

However, the computation of deep neural networks (DNN) is very resource heavy in energy, compute power and memory, in both space and bandwidth. These problems have been circumvented by moving the data from the generating

© Springer Nature Switzerland AG 2019
A. Monreale et al. (Eds.): ECML PKDD 2018 Workshops, CCIS 967, pp. 5–20, 2019.
https://doi.org/10.1007/978-3-030-14880-5_1

device at the edge to a centralized computation unit (i.e. cloud service). But as the number of devices and the demands for low latency increase, moving large amounts of data away from the device becomes infeasible. The training of deep networks on distributed embedded systems is even more demanding, as it requires to send updates of all weights between the workers.

A key observation is that a large portion of the parameters in a neural network are redundant. If the operations and activations that are not necessary can be identified, this can make the calculation more efficient and save energy. It has been shown that models can be compressed by a factor of over 30 [1].

1.1 Related Work

There have been several ideas on how to enable deep learning on the edge by removing redundancy. A very well received approach is to use topologies which are specifically designed to be very lean and thus avoid redundancy by design [2–4]. This approach has a high popularity for mobile applications.

Instead of looking at the efficiency problem from an algorithmic side, the hardware can be adapted to be very efficient for computations required by DNNs. The calculation of the operations in a layer can be moved to a co-processor.

An option are FPGAs, which allow to design a specialized hardware architecture for DNNs, but at much less effort than building a computer chip from scratch. There are several examples of FPGA implementations dealing with redundancy in DNNs [5–10]. FPGAs consume little energy, therefore they are good candidates for embedded applications.

Redundancy reduction can also be seen in the context of distributed systems. Training DNNs on such systems is an active field of research [11–13]. A problem is the transfer of the weight updates in the form of gradients between the different nodes. Redundancy appears in weight updates which have no effect on the convergence of the training. If that information can be prevented from being transferred, it can save bandwidth [14–17]. It has been shown that compression ratios of up to 600 in memory size are possible [18].

1.2 Contribution

All of the approaches above can be interpreted as a way to introduce and leverage sparsity in deep neural networks. Sparsity is the ratio of zero-value elements to all elements. This concept is applicable most directly to the weights in a neural network, but also reducing the number and shape of layers can be interpreted as a form of sparsity leveraging.

The results for novel methods found in literature often use small topologies and simple datasets as a proof of concept. It is not clear if those results transfer well to bigger models, i.e. if these methods scale. For hardware accelerators, which have to rely on a certain amount of sparsity in order to be efficient, it is crucial to know if a certain topology is able to deliver unchanged results with a sparse representation of the data. There is a need for a methodology that can investigate the potential in sparsity of a model prior to the hardware implementation.

This paper extends the capabilities of *TensorQuant*[1] [19], which is an open-source toolbox for *TensorFlow* [20]. It can be used to investigate sparsity in custom topologies and datasets with very little changes to the original files describing the model. The contributions in this paper are:

- Sparsity is studied in several convolutional neural network (CNN) topologies of varying sizes. The differences in the sparsity of the activations and weights during inference are investigated.
- The sparsity of the gradient during training is examined. This shows which level of accuracy can be expected for different gradient sparsity levels, if no additional methods are applied to guide the training process.
- *TensorQuant* is extended and used to provide an easy way to access and manipulate the layers in a DNN for sparsity experiments. It offers an open platform to test and compare various methods which rely on tensor alteration, including sparsity.

This work puts methods which leverage sparsity into perspective, as it shows what level of sparsity can already emerge from using regular methods.

Section 2 introduces the used terms and methods in this paper. It gives a brief overview of *TensorQuant* and how it can help to investigate sparsity. In Sect. 3 experiments are conducted, which show to which degree sparsity is emerging in CNNs, applying regular methods for training and inference.

2   Methods

A neural network layer is defined as

$$z_l = f(x_l, W_l), \tag{1}$$

where x_l is the input, z_l is the activation and W_l is a set of weights in the layer l. f is a non-linear function, called activation function. A neural network is trained by minimizing some loss function $L(W)$, which can include terms for L1 and L2 regularization [21]. The optimization step is

$$w_{t+1} = w_t - \lambda \frac{\partial L}{\partial w_t} \tag{2}$$

for every weight w in the neural network. $\frac{\partial L}{\partial w_t}$ is referred to as the gradient, and it is scaled with some learning rate λ before applied to the weight as an update.

---

[1] www.tensor-quant.org.

2.1 Sparsity

Sparsity is defined as the ratio of zero-value elements to all elements. The sparsity of a layer is

$$s_l = \frac{|\{w \mid w = 0, w \in W_l\}|}{|W_l|}. \tag{3}$$

In a large model comprising of many layers it helps to group layers in a logical hierarchy, referred to as a block

$$B = \{l \mid \text{for some arbitrarily chosen } l\}. \tag{4}$$

Then the sparsity of that block is the number of all zero weights divided by the number of weights belonging to that block

$$s_b = \frac{\sum_{l \in B} |W_l| s_l}{\sum_{l \in B} |W_l|}, \tag{5}$$

with B being the set of all layers belonging to the logical block. The total sparsity of a model can be calculated similarly, but summing over all layers in all blocks.

The gradients of the weights are grouped in the same way as the weights themselves and their sparsity is computed in the same manner.

The activation sparsity is different from the weights and gradients. It always refers to the last activation of a block, without considering the other layers within the block as in Eq. (5). It is defined similar to Eq. (3), but by counting over the activation values z instead of weights w.

2.2 Enforcing Sparsity

When using a ReLU as the activation function, sparsity emerges in the activations to a high degree. For the weights and gradients, however, it is very unlikely that even one of their values is exactly zero. Applying Eq. (3) will always result in zero sparsity, as the filters are, in fact, dense. Therefore, it is necessary to enforce sparsity. One method is to select a certain number of elements with the highest magnitude and set the other ones to zero [22]. Another way is to use a threshold for the magnitude and to set all the values below it to zero. The latter approach is used in this paper.

2.3 TensorQuant

TensorQuant is a toolbox for *TensorFlow*, originally designed to investigate the effects of quantization on deep neural networks [19]. One of its distinct features is that it can manipulate the tensors in a network to a very deep level, without much changes to the files describing the model. Manipulation is performed by looping in additional operations at specific locations. *TensorFlow* allows to introduce userdefined C++ kernels as additional tensor operators. In *TensorQuant*, those operations are referred to as quantizers. Thus, a quantizer or kernel can

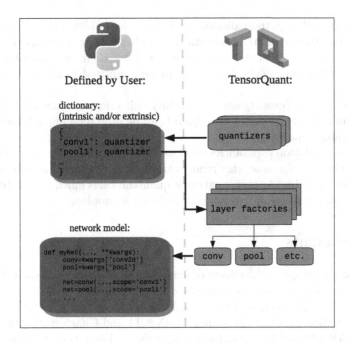

Fig. 1. Overview of the TensorQuant workflow. The user provides a python dictionary, which maps variable scopes to quantizer objects. Minor changes need to be applied to the file describing the topology, so that *TensorQuant* can loop in the quantizers at the desired locations.

be designed which sets all entries to zero whose magnitude is below a certain threshold. By incorporating this operation into a model, the weights, activations and gradients can be sparsified.

The lowest level where tensors can be manipulated in the context of quantization is referred to as intrinsic quantization. Every layer is broken down to its tensor operations. The tensors passing from one operation to the next are quantized at every step in order to assure that the precision of the intermediate result never exceeds the one of the data format to be emulated. This way, *TensorQuant* can emulate low-bitwidth operations, specifically in the convolution layers of CNNs.

Another location where tensor manipulation can be applied is just at the output of a layer. In the context of quantization, this is called extrinsic quantization, and it is the location where activation sparsification can be introduced. The weights can be manipulated just before they are passed to the operations, which allows to sparsify weights. Also the gradients can be sparsified before they are being applied to the weights as updates.

TensorQuant uses the *TensorFlow slim* framework to provide a variety of utility functions. *TensorQuant* extends this framework by adding several additional functionalities which ease the access to the layers. See Fig. 1 for an overview of the workflow. The layers in *TensorFlow* are tagged with so called variable scopes. If a python dictionary is provided which maps these scopes to quantizers, *TensorQuant* automatically applies tensor manipulation to the desired locations. Every layer and block can have their own quantizer. This automatism allows very deep topologies to be quantized easily with arbitrary granularity.

As for now, the *TensorQuant slim* utility collection is made for CNNs and classification tasks. However, *TensorQuant* can be used on a much broader class of deep learning topologies.

Using the emulation capabilities of *TensorQuant* comes at the cost of degrading the runtime. As for now, this renders training in combination with intrinsic quantization infeasible. Applying extrinsic quantization is much less problematic, so that training with sparsifying operators is not a problem.

3   Experiments

This section investigates the effects of sparsity enforcement on some CNN classifiers. The choice of topologies reflects different difficulty levels. AlexNet [23] and ResNet 50 [24] are trained on the ILSVRC12 ImageNet [25] dataset, which is the most difficult task in this paper. ResNet 14 and CifarNet [26] are trained on the CIFAR 100 and 10 [27] dataset, respectively. These two are considered to be medium and low level problems. Finally, the MNIST dataset is used to train LeNet [28], which is considered to be a trivial problem.

AlexNet and CifarNet use dropout [29], whereas the Inception and ResNet topologies use batch normalization [30]. All topologies use ReLUs as activation functions. These special layers can have an additional impact on the sparsity, but which is not investigated in this paper.

The naming convention of the layers is as follows: "conv" refers to a single convolution layer, "logits" and "fc" to fully connected ones. In the ResNet topologies, a "block" is a logical block comprising of convolution layers with the same number of input and output filters. A "unit" comprises of a bottleneck layer [24], which has three convolutions plus a shortcut connection.

3.1   Sparsity of Activations and Weights

If the weights of a DNN model are sparse, the required memory to store the model can be decreased. A high sparsity in activation values can decrease the computation time, even if the weights are not sparse. Therefore it is interesting to look at the sparsity of weights and activations. Normally, a L2 regularizer is used during training. It is known that a L1 regularizer promotes sparsity in the weights, although it makes convergence to an optimum more difficult and therefore it is less often used. This section shows the different sparsity levels between a variety of CNN topologies, trained with L1 and L2 regularizers, respectively. The focus of this section is on the inference.

First, the network is trained without any sparsity enforcement with either L1 or L2 regularization. As explained in Sect. 2.2, weights are not sparse as they are, so sparsity needs to be enforced, e.g. with thresholding. The objective here is to set as many weights to zero without retraining, so that the test accuracy is not changed. To obtain a high total sparsity, each layer or block has its own threshold. They are found with a grid search approach, by going through all layers or blocks iteratively. For each layer, the highest threshold is found which leaves the test accuracy unchanged. Meanwhile, the other layers are not sparsified. For the test accuracy, all thresholds for all layers are applied at once. The values for the test accuracies are stated in the captions of the respective figures, relative to the L2 test accuracy without sparsification.

Although activations could be sparsified with the same method, it has proven to be rather ineffective in our experience. The sparsity which comes from the ReLU activation functions is already high and further thresholding does not have much effect.

Table 1. LeNet weight and activation sparsity after training with L1 and L2 regularizer. The relative test accuracies are L1 99.0% and L2 99.8%.

Layer	L2 weights	L1 weights	L2 activations	L1 activations
conv1	0.142	0.289	0.717	0.513
conv2	0.491	0.505	0.528	0.662
fc3	0.258	0.502	0.661	0.593
fc4	0.258	0.504	0.000	0.000

Table 1 shows the weight and activation sparsities for LeNet trained with L1 and L2 regularization, respectively. L1 increases the sparsity of weights in every layer, especially in the last two fully connected layers. The activations change in sparsity as well, though there is no general trend. The last layer is the classification output, so it is no surprise that the sparsity is zero. Notice that in LeNet the test accuracy is higher with L1 regularization than for L2.

Table 2. CifarNet weight and activation sparsity after training with L1 and L2 regularizer. The relative test accuracies are L1 101.8% and L2 98.4%.

Layer	L2 weights	L1 weights	L2 activations	L1 activations
conv1	0.064	0.117	0.683	0.665
conv2	0.226	0.754	0.854	0.837
fc3	0.164	0.623	0.781	0.797
fc4	0.563	0.501	0.649	0.458
logits	0.862	0.120	0.000	0.000

CifarNet in Table 2 shows a more dramatic increase in weight sparsity in some of its layers. But surprisingly, the sparsity in the last layer dropped a lot. The activation sparsities, however, are mostly unchanged between L1 and L2.

Table 3. ResNet 14 weight and activation sparsity after training with L1 and L2 regularizer. The relative test accuracies are L1 93.2% and L2 99.6%.

Layer	L2 weights	L1 weights	L2 activations	L1 activations
conv1	0.051	0.382	0.285	0.273
block1/unit_1	0.188	0.390	0.428	0.407
block1/unit_2	0.074	0.501	0.274	0.257
block2/unit_1	0.058	0.313	0.673	0.699
block2/unit_2	0.056	0.349	0.766	0.745
logits	0.028	0.233	0.000	0.000

Table 4. AlexNet weight and activation sparsity after training with L2 regularizer. The relative test accuracy is L2 98.0%.

Layer	L2 weights	L2 activations
conv1	0.161	0.604
conv2	0.177	0.804
conv3	0.177	0.825
conv4	0.406	0.863
conv5	0.219	0.920
fc6	0.524	0.817
fc7	0.730	0.838
fc8	0.474	0.000

The ResNet 14 topology in Table 3 exhibits a very low weight sparsity for L2 and rises only moderately with L1. The activation sparsity does not change between L1 and L2, as it was the case with CifarNet.

Training AlexNet with a L1 regularizer is difficult, and even when incorporating a mixed L1-L2 regularization, the results remain poor, so only the results for L2 regularization are shown. The L2 weight sparsity for AlexNet in Table 4 is low for most of the convolution layers, but high for the fully connected ones. The activation sparsity is very high in all layers.

ResNet 50 also trains poorly with the L1 regularizer. For L2, Table 5 shows that the weight sparsity is not high for most of the layers. Activation sparsity is rather low. The "unit 1" layers in every block have the highest activation sparsity, except for the layers in block 4, where it is very high in all units.

Table 6 gives an overview of the total weight sparsities for L1 and L2 regularization. It shows that for simpler problems, it is easy to achieve high sparsity even with simple regularization methods. In all topologies, the weight sparsity is lower than the one for activations, which agrees with observations made in other work [5,17,31,32]. Identifying layers with sparse activations contains valuable information for model parallelism. They are good locations to cut the topology

Table 5. ResNet 50 weight and activation sparsity after training with L2 regularizer. The relative test accuracy is L2 99.6%

Layer	L2 weights	L2 activations
conv1	0.073	0.301
block1/unit_1	0.378	0.359
block1/unit_2	0.369	0.236
block1/unit_3	0.359	0.214
block2/unit_1	0.122	0.493
block2/unit_2	0.489	0.343
block2/unit_3	0.126	0.304
block2/unit_4	0.125	0.268
block3/unit_1	0.177	0.542
block3/unit_2	0.344	0.417
block3/unit_3	0.178	0.369
block3/unit_4	0.370	0.336
block3/unit_5	0.391	0.306
block3/unit_6	0.210	0.270
block4/unit_1	0.242	0.735
block4/unit_2	0.510	0.737
block4/unit_3	0.265	0.824
logits	0.127	0.000

Table 6. Overview of total weight sparsity after training with L1 and L2 regularizer.

Layer	L2	L1
LeNet	0.262	0.502
CifarNet	0.185	0.624
ResNet 14	0.062	0.326
AlexNet	0.573	–
ResNet 50	0.291	–

into subgraphs, which can be put on separate nodes in a distributed system. This allows to minimize the amount of data which needs to be transferred. For instance, the "unit 1" layers of each block in ResNet 50 would be good separation points.

3.2   Sparsity of Gradients During Training

When training on a distributed system, the sparsity in the gradients can help to reduce the amount of data which needs to be transferred to compute an update.

So in this section, the sparsity of the gradient is investigated during training. Similar to the weights, sparsity needs to be enforced. In a single training run, the same threshold is applied to all gradients in every step. L2 regularization is used during training.

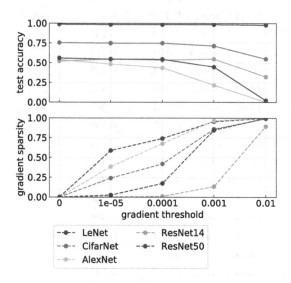

Fig. 2. Comparison of absolute test accuracy versus the applied gradient threshold during training.

Figure 2 shows the final test accuracy and the total gradient sparsity towards the end of the training versus the applied gradient threshold. All training runs have the same number of iterations as the baseline. A threshold of zero indicates the baseline. AlexNet is more susceptible for sparsified gradients than the ResNet topologies. For ResNet14, the sparse gradients have a regularizing effect, so the test accuracy increases above the baseline if the threshold is not too high. Such a regularizing effect has also been observed in other work [22]. CifarNet is mostly unchanged similarly to ResNet 14, but does not show the same regularizing effect. The LeNet topology is almost unchanged for the investigated thresholds.

The gradient sparsity can be almost 100% for LeNet and the model can still learn fine. This indicates that MNIST on LeNet is a rather trivial problem. AlexNet shows a steady decline in test accuracy with an increasing threshold. CifarNet and the two ResNet architectures have a jump in gradient sparsity, but where the test accuracy does not change much. ResNet 50 can achieve 80% baseline accuracy at a gradient sparsity of 85%. This suggests that there is a sweet spot for the gradient threshold, which allows for very high sparsities in those topologies.

Figures 3, 4, 5, 6 and 7 show how the gradient sparsity evolves during training for individual layers or blocks. The weights are initialized with a gaussian distribution. AlexNet, LeNet and CifarNet are trained with a batch size of 128, and 32 for the ResNet topologies. The gradient thresholds are chosen in such a way that the final test accuracy is close to the baseline accuracy, but also that there is some visible gradient sparsity. The gradient thresholds and achieved accuracies are stated in the captions of the figures.

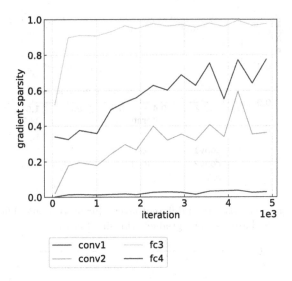

Fig. 3. LeNet gradient sparsity of different layers during training. The achieved test accuracy is 99% of the baseline. The gradient threshold is 10^{-3}.

The layerwise gradient sparsity for LeNet is shown in Fig. 3. The gap between the first layer and fc3 is striking. This suggests that conv1 holds the most information, whereas fc3 is redundant.

The evolution of the gradient sparsities in CifarNet differ somewhat from LeNet, even though the topologies are very similar. There is a very high peak in sparsity at a very early phase of the training, except for the logits layer. Each layer seems to converge towards a certain sparsity level, where the fc3 layer, which is in the middle of the topology, has the highest sparsity.

Resnet 14 in Fig. 5 comprises almost entirely of convolution layers, only the last layer is fully connected. All gradient sparsities increase rapidly in the first epoch, then they show a decreasing trend. The logits layer has a higher sparsity than all the other layers. Different to the two topologies before, the convolution layers exhibit a similar, low sparsity. The decreasing trend in gradient sparsity seems to contradict the fact that the gradient is becoming flatter the closer the weights converge to an optimum. However, the decrease in gradient sparsity only means that the number of non-zero elements is increasing, it does not imply anything about the magnitude of the gradient itself. A possible explanation is that the gradient is pointing more equally into multiple dimensions when it gets closer to an optimum than at the beginning of the training.

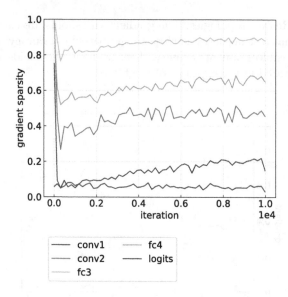

Fig. 4. CifarNet gradient sparsity of different layers during training. The achieved test accuracy is 97% of the baseline. The gradient threshold is 10^{-3}.

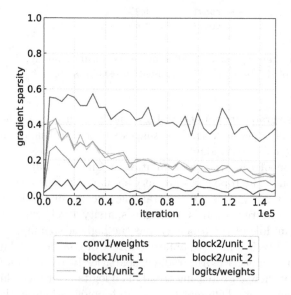

Fig. 5. ResNet 14 gradient sparsity of different layers during training. The achieved test accuracy is 105% of the baseline. The gradient threshold is 10^{-3}.

For AlexNet in Fig. 6, the fully connected layers show a more pronounced gap to the convolution layers than the topologies before. The fully connected layers also have a more unique behavior. There is an initial decline in sparsity in the beginning, followed by an increase after a few epochs. Then, the gradient

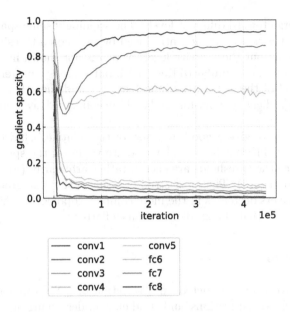

Fig. 6. AlexNet gradient sparsity of different layers during training. The achieved test accuracy is 80% of the baseline. The gradient threshold is 10^{-4}.

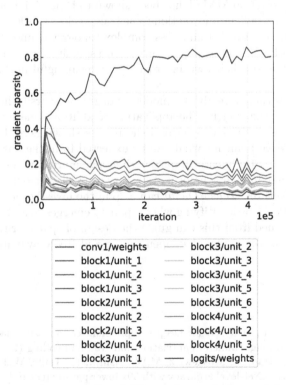

Fig. 7. ResNet 50 gradient sparsity of different layers during training. The achieved test accuracy is 96% of the baseline. The gradient threshold is 10^{-4}.

sparsities converge toward different levels. The seemingly same sparsity level for the convolution layers is a artifact introduced by the chosen threshold. A higher value would spread out the sparsity levels, which is not shown here.

ResNet 50 in Fig. 7 is similar to ResNet 14, but there is a bigger gap between the sparsity level of the last, fully connected layer and the other convolutional ones. Also, the gradient sparsity for the last fully connected layer goes up instead of down.

The figures above give a good overview of the relative behavior of the gradient sparsities of different layers. The absolute values of the sparsities are less meaningful, since the thresholds are chosen rather arbitrarily (Fig. 2 is a better reference in that regard). The most striking result is that the gradient sparsity of convolution layers decreases in the more complex topologies, which seemingly contradicts the fact that the gradient becomes flatter.

4   Conclusion

Experiments have been conducted on a selection of CNN topologies, showing sparsity for weights, activations and gradients under changing problem size. Although all of them are CNN classifiers, there are differences in where and to which degree sparsity emerges, especially in the gradients during training. The training of LeNet on MNIST has been shown to be a trivial problem, which requires almost no gradient information to be trained close to 100% test accuracy. Therefore, results obtained from a less complex topology cannot be transferred to deeper networks. It is necessary to investigate sparsity for each topology and sparsifying method on their own in order to get meaningful information about sparsity.

In many cases there already is a moderate degree of sparsity in the regularly trained versions of the models. The application of additional methods to promote sparsity can increase the levels beyond the results shown here, but this paper serves as a reference point for what can be expected from the baseline model.

Our results back the idea of implementing sparse arithmetics on embedded devices, since the redundancy in form of sparsity can be leveraged through special hardware architectures. *TensorQuant* can help in the investigation of sparsity in deep neural networks by identifying where sparsity emerges to a high degree. The information obtained from this can guide the design of sparse arithmetics hardware accelerators. *TensorQuant* is open-source and freely available on *GitHub* (See footnote 1).

References

1. Han, S., Mao, H., Dally, W.J.: Deep compression: compressing deep neural networks with pruning, trained quantization and Huffman coding (2015)
2. Iandola, F.N., Han, S., Moskewicz, M.W., Ashraf, K., Dally, W.J., Keutzer, K.: SqueezeNet: AlexNet-level accuracy with 50x fewer parameters and <0.5 MB model size (2016)

3. Howard, A.G., et al.: MobileNets: efficient convolutional neural networks for mobile vision applications (2017)
4. Zhang, X., Zhou, X., Lin, M., Sun, J.: ShuffleNet: an extremely efficient convolutional neural network for mobile devices (2017)
5. Zhu, J., Jiang, J., Chen, X., Tsui, C.Y.: SparseNN: an energy-efficient neural network accelerator exploiting input and output sparsity. CoRR, abs/1711.01263 (2017)
6. Han, S., et al.: EIE: efficient inference engine on compressed deep neural network. In: ISCA, pp. 243–254. IEEE Computer Society (2016)
7. Aimar, A., et al.: NullHop: a flexible convolutional neural network accelerator based on sparse representations of feature maps (2017)
8. Andri, R., Cavigelli, L., Rossi, D., Benini, L.: YodaNN: an architecture for ultra-low power binary-weight CNN acceleration (2016)
9. Rybalkin, V., Wehn, N., Yousefi, M.R., Stricker, D.: Hardware architecture of bidirectional long short-term memory neural network for optical character recognition. In: Proceedings of the Conference on Design, Automation and Test in Europe, pp. 1394–1399. European Design and Automation Association (2017)
10. Chang, A.X.M., Zaidy, A., Gokhale, V., Culurciello, E.: Compiling deep learning models for custom hardware accelerators (2017)
11. You, Y., Gitman, I., Ginsburg, B.: Large batch training of convolutional networks (2017)
12. Keuper, J., Pfreundt, F.J.: Distributed training of deep neural networks: theoretical and practical limits of parallel scalability (2016)
13. Kuehn, M., Keuper, J., Pfreundt, F.J.: Using GPI-2 for distributed memory paralleliziation of the Caffe toolbox to speed up deep neural network training (2017)
14. Renggli, C., Alistarh, D., Hoefler, T.: SparCML: high-performance sparse communication for machine learning (2018)
15. Aji, A.F., Heafield, K.: Sparse communication for distributed gradient descent (2017)
16. Wangni, J., Wang, J., Liu, J., Zhang, T.: Gradient sparsification for communication-efficient distributed optimization (2017)
17. Rhu, M., O'Connor, M., Chatterjee, N., Pool, J., Kwon, Y., Keckler, S.W.: Compressing DMA engine: leveraging activation sparsity for training deep neural networks. In: 2018 IEEE International Symposium on High Performance Computer Architecture (HPCA), pp. 78–91. IEEE (2018)
18. Lin, Y., Han, S., Mao, H., Wang, Y., Dally, W.J.: Deep gradient compression: reducing the communication bandwidth for distributed training (2017)
19. Loroch, D.M., Pfreundt, F.J., Wehn, N., Keuper, J.: TensorQuant: a simulation toolbox for deep neural network quantization. In: MLHPC@SC, pp. 1:1–1:8. ACM (2017)
20. Abadi, M., et al.: TensorFlow: a system for large-scale machine learning. OSDI 16, 265–283 (2016)
21. Bottou, L., Curtis, F.E., Nocedal, J.: Optimization methods for large-scale machine learning. SIAM Rev. 60(2), 223–311 (2016)
22. Sun, X., Ren, X., Ma, S., Wang, H.: meProp: sparsified back propagation for accelerated deep learning with reduced overfitting. CoRR, abs/1706.06197 (2017)
23. Krizhevsky, A., Sutskever, I., Hinton, G.E.: ImageNet classification with deep convolutional neural networks. In: Advances in Neural Information Processing Systems, pp. 1097–1105 (2012)
24. He, K., Zhang, X., Ren, S., Sun, J.: Deep residual learning for image recognition (2015)

25. Deng, J., Dong, W., Socher, R., Li, L.J., Li, K., Fei-Fei, L.: ImageNet: a large-scale hierarchical image database. In: IEEE Conference on Computer Vision and Pattern Recognition, CVPR 2009, pp. 248–255. IEEE (2009)
26. Krizhevsky, A.: Learning multiple layers of features from tiny images, May 2012
27. Krizhevsky, A., Nair, V., Hinton, G.: CIFAR-100 (Canadian institute for advanced research) (2009)
28. LeCun, Y., Bottou, L., Bengio, Y., Haffner, P.: Gradient-based learning applied to document recognition. Proc. IEEE **86**(11), 2278–2324 (1998)
29. Srivastava, N., Hinton, G.E., Krizhevsky, A., Sutskever, I., Salakhutdinov, R.: Dropout: a simple way to prevent neural networks from overfitting. J. Mach. Learn. Res. **15**(1), 1929–1958 (2014)
30. Ioffe, S., Szegedy, C.: Batch normalization: accelerating deep network training by reducing internal covariate shift. CoRR, abs/1502.03167 (2015)
31. Wen, W., Wu, C., Wang, Y., Chen, Y., Li, H.: Learning structured sparsity in deep neural networks (2016)
32. Liu, X., Pool, J., Han, S., Dally, W.J.: Efficient sparse-Winograd convolutional neural networks. CoRR, abs/1802.06367 (2018)

Asynchronous Federated Learning
for Geospatial Applications

Michael R. Sprague$^{(\boxtimes)}$ (iD), Amir Jalalirad (iD), Marco Scavuzzo (iD),
Catalin Capota (iD), Moritz Neun (iD), Lyman Do (iD), and Michael Kopp (iD)

HERE Technologies, Amsterdam, The Netherlands
michael.sprague@here.com
http://www.here.com

Abstract. Federated learning is an emerging collaborative machine-learning paradigm for training models directly on *edge devices*. The data remains on the edge device and this method is robust under real-world edge data distributions. We present a new asynchronous federated-learning algorithm ('asynchronous federated learning') and study its convergence rate when distributed across many edge devices, with hard data constraints, relative to training the same model on a single device. We compare asynchronous federated learning to an existing synchronous method. We evaluate its robustness in real-world situations; for example, devices joining part-way through training or devices with heterogeneous compute resources. We then apply asynchronous federated learning to a challenging geospatial application, namely image-based geolocation using a state-of-the-art convolutional neural network. Our results lay the groundwork for deploying large-scale federated learning as a tool to automatically learn, and continually update, a machine-learned model that encodes location.

Keywords: Federated learning · Asynchronous communication ·
Heterogeneous computation · Image-based geolocation

1 Introduction

In the currently prevailing approach to machine learning on *edge devices*, where sensor data is collected, the training data is transferred to a data centre, where the models are trained and housed. During model inference, the input data is transferred to the data centre, and the prediction is sent back to the edge device collecting the data. An alternative approach is to perform inference directly on the edge device [2]. Although this approach has the advantage of reducing latency, cost, and bandwidth, it requires sufficient input data to be transferred in order to train the model. Furthermore, shifts in the underlying data distribution via *concept drift* or learning from rare outliers requires a process to continually collect data, retrain the model, and upload the model to the edge device. The ongoing transfer of user data poses a significant risk to users' privacy.

© Springer Nature Switzerland AG 2019
A. Monreale et al. (Eds.): ECML PKDD 2018 Workshops, CCIS 967, pp. 21–28, 2019.
https://doi.org/10.1007/978-3-030-14880-5_2

This privacy risk has motivated work to learn directly on edge devices without transferring the data off the device. However, a model trained on a single edge device is unlikely to effectively generalize as the data collected on the edge is non-independent and identically distributed (non-i.i.d.) [7]. Worse, the distribution of data is likely unbalanced and, thus, devices may have orders of magnitude different data at their disposal [7].

An emerging solution, *federated learning*, allows for input data to stay on each edge device but pools the combined knowledge across devices regularly [7]. It is stable under non-i.i.d. and unbalanced input data distributions and can be applied to a wide class of learning algorithms train-able via gradient descent [3,6,7]. A recently proposed algorithm, *federated averaging* [7], synchronously aggregates the parameters during training via a *parameter server* to learn a communal model while minimizing the number of communication rounds [7]. The parameters are weighted by the number of data points per device. For each communication round, a sub-sample of devices participates in the round. The communication round proceeds as fast as the slowest device and, as a result, this approach is susceptible to the *straggler effect*. Furthermore, the impact of constraining data on the edge device, without shuffling data between devices as in a data centre, has not been explicitly evaluated. And real-world effects were not considered, such as devices joining mid-way through training or devices with heterogenous compute resources. Finally, extending federated learning beyond relatively small neural networks has not been demonstrated.

Our Contribution. In order to handle real-world scenarios where devices randomly leave or rejoin the training process, and where the update rates of edge devices can vary substantially, we ask whether the synchronous condition of the federated learning algorithms can be relaxed. This builds on previous work on asynchronous gradient descent [5]. Specifically, we make these contributions:

1. We examine the performance degradation caused by imposing hard data constraints on edge devices compared to the case where the training occurs on a single device.
2. We propose an asynchronous aggregation scheme for federated learning. We compare its performance to the baseline synchronous federated-averaging algorithm [7]. We show its viability and robustness to real-world situations where devices join partway through training or train at different speeds.
3. We present an application of federated learning for end-to-end localization, without an explicit map, using a state-of-the-art convolutional neural network.

Algorithm. We consider a supervised-learning problem with the data partitioned onto k edge devices $B_1, \cdots, B_k$. Each device fetches the latest global parameter, θ_n^g and updates its local parameters θ_n^k at time $n+1$ by training on E local epochs with learning rate α and, $\forall k$,

$$\theta_{n+1}^k \leftarrow \theta_n^g - \alpha \frac{\partial h_k}{\partial \theta_n^k}\bigg|_{\theta_n^g} \tag{1}$$

where $h_k(x, y, \theta)$ is the model loss function for input x and label y.

After the local training, which can involve multiple epochs of the local data, the edge device's parameters are sent to the central parameter server. The parameter server then immediately aggregates the model parameters without waiting for any other edge devices and then returns the aggregated parameters to the device

$$\theta_{n+1}^g \leftarrow (1 - \eta^k)\theta_n^g + \eta^k \theta_{n+1}^k \qquad (2)$$

where aggregation weight η^k, in general, is a function of device-specific meta-parameters (number of data points, time since last update, local training loss, etc.). We found that, with the datasets examined here, a constant aggregation weight on the order of 0.1, independent of the number of data points per device, struck the right balance between stability and convergence speed. Note there are no limits on the relative contribution of the different edge devices. Nor is it clear what conditions need to be satisfied to guarantee convergence. Surprisingly, we demonstrate below that this update mechanism does converge and, furthermore, that it is robust not only to non-i.i.d. and unbalanced data distributions across devices, but it can also handle realistic device failures.

2 Results

Experimental Setup. We initially evaluated the algorithms on two datasets, MNIST and CIFAR-10. We used two two-layer convolutional neural networks followed by two fully-connected layers (total parameters 169 510 and 576 790, respectively). To simulate a non-i.i.d. distribution on the edge devices, we partitioned the data such that there were only two unique data labels per device [7]. The data points for edge device k were sampled from a symmetric Dirichlet distribution, parameterized by a concentration term λ, and with marginal distribution $X_k \sim \text{Beta}(\lambda, K\lambda - \lambda)$ where K is the number of edge devices. When $\lambda \to \infty$, the distribution tends to $1/K$ and, when $\lambda \to 0$, all the probability mass is concentrated in one device.

We simulated 100 edge devices communicating with a single parameter server. Edge devices were developed with *Torch/PyTorch*. At start up, each device downloaded a disjoint partition of the data. An additional edge device functioned as a *test device* that regularly checked the accuracy of the global model against the test dataset. All of the components sent their metrics to a shared storage system in the cloud. Upon experiment termination, an Apache Drill-based tool automatically summarized the results of the experiment. We ran the components as *Docker containers* on an AWS cluster composed of one r4.16xlarge virtual machine (VM), which we call the *master node*, and 30 r4.2xlarge VMs, which we call *worker nodes*. We used *Docker Swarm* to allocate the containers and assign memory constraints to the edge devices. In particular, the master node executed the *swarm manager* and hosted an instance of the parameter server and a test-device container. Each worker node ran up to four edge-device containers in parallel.

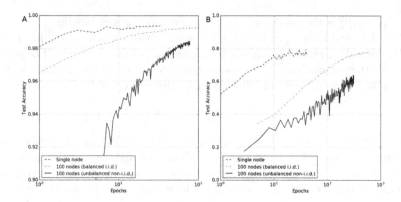

Fig. 1. Comparison of single-node training (all data on one device, with shuffling between epochs) and data distributed on edge devices with no data shuffling between epochs. For the unbalanced experiments, $\lambda = 1$. A. MNIST dataset B. CIFAR-10 dataset

Convergence Relative to Single-Node Training. We examined the convergence rates of training with asynchronous federated learning. Figure 1 compares the number of epochs as a function of test accuracy using the standard test sets defined for the MNIST and CIFAR-10 datasets. In the multi-node experiments, the ordering of device updates is not controlled and an epoch here is defined as the sum of the number of data points on a device each time a device communicates with the parameter server, divided by the total number of data points. For the MNIST dataset, the number of epochs needed to reach 98% accuracy increased by a factor 4.5 ± 0.4 and 55 ± 5 for the balanced, i.i.d. and unbalanced non-i.i.d. distributions, respectively. For the CIFAR-10 dataset, the number of epochs needed to reach 60% accuracy increased by a factor ~10 and ~120 for the balanced i.i.d. and unbalanced non-i.i.d. distributions, respectively.

Comparing Synchronous vs Asynchronous Schemes. In Fig. 2, we compare the test accuracy as a function of the number of communication rounds and wall-clock time for the asynchronous and synchronous aggregation schemes using the MNIST dataset. Each of the simulated edge devices were provisioned with the same resources. The hyper-parameters that affect the runtime (e.g. batch size, local epochs processed) were held jointly constant. For the balanced, i.i.d. and non-i.i.d. data, the performance was similar in terms of communication rounds and wall-clock time. For the balanced, non-i.i.d. data, the increased variance of the test accuracy shows the tension between local and global optimization of the model parameters. For the unbalanced, non-i.i.d. data, the edge devices with fewer data points communicated more often with the parameter server; the local processing time was linearly proportional to the number of data points. This resulted in factor of 5 ± 1 decrease in wall-clock time to reach 95%

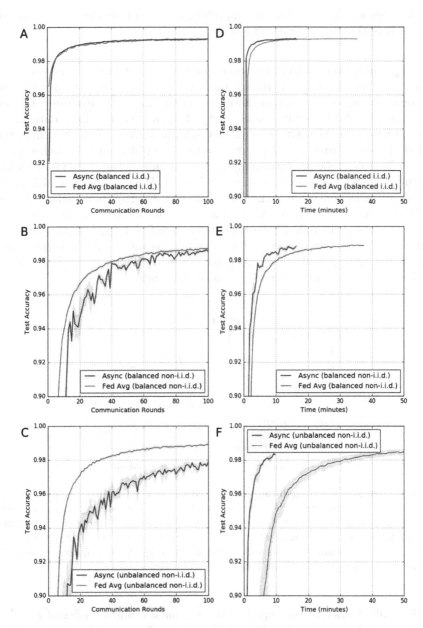

Fig. 2. Asynchronous vs synchronous federated learning for different data distributions as a function of communication rounds (defined as any 100 devices communicating once) and wall-clock time. The grey shaded area shows the standard error from three repetitions with data sampled from the underlying data distribution drawn from MNIST. The asynchronous aggregation weighting was 0.1, batch size was 10, local epochs (E) was 5, learning rate (with vanilla SGD) was 0.05.

test accuracy for the asynchronous approach, in spite of the slower convergence as a result of the skewed update rates.

Heterogenous Device Performance. In a realistic deployment of federated learning, devices may join part-way through a training run, or edge devices may process data locally at substantially different rates. We examined this by partitioning the data into two groups of edge devices, each containing only half of the total labels. At the start, 50 edge devices with half the labels (0–4) from the MNIST dataset started training. As expected, the test accuracy reached close to 50%. At either 1000 or 7000 communication rounds, 50 edge devices with the remaining labels (5–9) joined, as shown in Fig. 3A. For the balanced, i.i.d. data, the test accuracy converged immediately, albeit with significant high-frequency oscillations in the test accuracy. For the unbalanced, non-i.i.d. data, the test accuracy 'plateaued', only increasing after a significant (and highly variable) delay when devices joined after 1000 communication rounds. In the experiments where the devices joined after 7000 communication rounds, the test accuracy remained at ~50% after 15500 communication rounds.

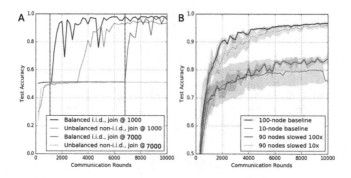

Fig. 3. A. New labels joining during training, either after 1000 communication rounds (blue dashed line, blue and green curves) or after 7000 communication rounds (red dashed line, red and cyan curves) with the MNIST dataset. B. Robustness of results to communication delays, where 90 of 100 edge devices (nodes) are slowed down by a factor of 10 or 100 compared to a baseline of all 100 edge devices running or only 10 edge devices running on the MNIST dataset. (Color figure online)

Figure 3B shows that the asynchronous scheme is robust to different processing speeds of the edge devices. Out of 100 devices, 90 were delayed by a factor of 10 or 100, which was compared to a baseline set of experiments where 100 devices ran normally, and an experiment where only 10 out of 100 edge devices (i.e. 90 devices have a delay that tends to infinity). The delayed-processing experiments converged between these two baselines, irrespective of the frequent and disruptive 'stale' parameter updates.

Image-Based Geolocation via Edge Learning. Finally, we evaluated the feasibility of applying asynchronous federated learning to a challenging geospatial application: image-based geolocation. Here we predict the camera's position and orientation from a set of images using a large convolutional neural network. We modified DenseNet121 [1], pre-trained on the ImageNet-1k dataset, with a 2048-neuron regression module [4]; the number of parameters is ~4 times smaller than the Inception-v3 architecture and the DenseNet model has lower error. The training and test images are from the King's College dataset [4]. The training data was randomly partitioned into 15 edge devices, with about 80 images per device. Quantitatively, we see the same behaviour as with the MNIST and CIFAR-10 datasets, where the convergence rate is slightly slower than on a single device, but the results eventually converge to similar accuracy as a non-federated model—here the minimum mean position error on the test data was 1.87 m (Fig. 4).

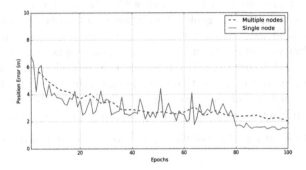

Fig. 4. Image-based geolocation of the King's College dataset where the data was randomly partitioned into multiple simulated edge devices (15 nodes/edge devices) and where all the data was partitioned on a single device (solid blue line). The learning rate was 10^{-4}, batch size was 16, momentum 0.9, 5 local epochs processed, ADAM used as the optimizer, and the asynchronous aggregation weighting was 0.2. (Color figure online)

3   Discussion and Outlook

We compared federated learning to training on a single-device without data constraints. We note that the slower convergence rate, in terms of epochs, is overcome, in terms of wall-clock time, by the parallelism inherent in federated learning. We did a detailed comparison between federated synchronous and asynchronous learning, where we found similar performance for balanced datasets with edge devices operating at the same speed. With the unbalanced dataset, the wall-clock performance of the asynchronous algorithm for the MNIST dataset was superior. Note that the specific performance is dependent on the non-linear scaling of the model accuracy vs the size of the dataset and any correlation between the update rate and data distribution. Further work is required to develop a principled approach for optimal aggregation based on desired optimization metrics (wall-clock time, local processing time, communication rounds), data and device-profile distributions, and additional resources like back-up workers.

We found that the asynchronous scheme had a graceful mode of failure; as the devices slowed down, the performance transitioned from a baseline of all devices operating normally to a baseline of only a subset of devices participating in training. The stale gradient updates from slow devices did not perturb the global parameter set sufficiently to prevent training. In contrast, edge devices with a subset of labels joining part-way through training did negatively affect the model convergence when the data was unbalanced, non-i.i.d. Interestingly, the test accuracy significantly lagged the moment when new devices joined. The increase in representational capacity of the models appears highly non-linear, akin to a phase transition for edge devices with non-i.i.d. data.

Finally, the image-based geolocation results demonstrate that federated learning is possible with a state-of-the-art convolutional neural network. We observed qualitatively similar trends in terms of convergence rate to the simpler MNIST and CIFAR-10 datasets. Taken together, federated learning represents a novel, scalable way to learn a location embedding on the 'edge'. Training models on edge devices such as autonomous cars is a possible route to the creation of a machine-learned, continuously updated representation of the world; a map.

Disclosure. The authors disclose they have filed a patent based on this work.

References

1. Huang, G., Liu, Z., van der Maaten, L., Weinberger, K.Q.: Densely Connected Convolutional Networks, August 2016. https://arxiv.org/abs/1608.06993
2. Iandola, F.N., Han, S., Moskewicz, M.W., Ashraf, K., Dally, W.J., Keutzer, K.: SqueezeNet: AlexNet-level accuracy with 50x fewer parameters and <0.5MB model size. arXiv:1602.07360 [cs], February 2016
3. Jiang, Z., Balu, A., Hegde, C., Sarkar, S.: Collaborative Deep Learning in Fixed Topology Networks. arXiv:1706.07880 [cs, stat], June 2017
4. Kendall, A., Grimes, M., Cipolla, R.: PoseNet: A Convolutional Network for Real-Time 6-DOF Camera Relocalization. arXiv:1505.07427 [cs], May 2015
5. Keuper, J., Pfreundt, F.J.: Asynchronous Parallel Stochastic Gradient Descent - A Numeric Core for Scalable Distributed Machine Learning Algorithms. arXiv:1505.04956 [cs], May 2015
6. Konečný, J., McMahan, H.B., Ramage, D., Richtárik, P.: Federated Optimization: Distributed Machine Learning for On-Device Intelligence. arXiv:1610.02527 [cs], October 2016
7. McMahan, H.B., Moore, E., Ramage, D., Hampson, S., Arcas, B.A.Y.: Communication-Efficient Learning of Deep Networks from Decentralized Data. arXiv:1602.05629 [cs], February 2016

Generalizing Knowledge in Decentralized Rule-Based Models

Pedro Strecht$^{(\boxtimes)}$ (iD), João Mendes-Moreira (iD), and Carlos Soares (iD)

INESC TEC/Faculdade de Engenharia, Universidade do Porto,
Rua Dr. Roberto Frias, 4200-465 Porto, Portugal
{pstrecht,jmoreira,csoares}@fe.up.pt

Abstract. Knowledge generalization of ruled-based models, such as decision trees or decision rules, have emerged from different backgrounds. This particular kind of models, given their interpretability, offer several possibilities to be combined. Despite each distinct context, common patterns have emerged revealing the systemic nature of the problem. In this paper, we look at the problem of generalizing the knowledge contained in a set of models as a process formalizing the operations that can be addressed in alternative ways. We also include a set-up to evaluate generalized models based on their ability to replace the base ones from a predictive performance perspective, without loss of interpretability.

Keywords: Knowledge generalization · Rule-based models

1 Introduction

Rules are usually presented in the canonical form of IF *antecedent* THEN *consequent*. The antecedent is a conjunction of relational conditions implicating independent variables to predict the value of a target variable of interest, the consequent. Rule-based models [14] make use of a set of rules to describe how independent variables can explain the value of an objective variable. A popular example are decision trees [9] which offer a flowchart representation of rules promoting easier human interpretation. Another are decision lists [13] which present ordered rule-sets in the canonical form. Interpretability is an important property in domains where a decision support system is able to explain and justify its decisions [7]. Therefore, the number of organizations using rule-based models has been increasing.

Generating models to predict or describe a phenomenon in organizations with a decentralized activity presents challenges. An example is a company that does its sales through subsidiaries or even by authorized individual distributors. Each sale is carried out by a single subsidiary which is considered a business unit of the organization. Another is of a university offering numerous courses to its students. Each course is offered by a faculty, or a department, which are further examples of business units. These organizations have their problem domain broken down into what can be seen as several units, i.e., a decentralized

© Springer Nature Switzerland AG 2019
A. Monreale et al. (Eds.): ECML PKDD 2018 Workshops, CCIS 967, pp. 29–36, 2019.
https://doi.org/10.1007/978-3-030-14880-5_3

context. Such parallelism makes it increasingly common to generate not a single model but multiple models, each relating to a business unit. In the company example, each subsidiary can have a model to describe/predict its monthly sales level. In the university context, each course can have a model to describe/predict the performance of the students enrolled in it. Yet, the fact that these models are associated with only one unit makes it hard to find generalized knowledge representative of the whole organization. In the aforementioned examples, this could be the overall monthly sales level behavior of the organization or the overall performance behavior of the students of the university during an academic year.

In this paper, we look at the problem of how to gather and generalize the knowledge contained in a large number of rule-based models from organizations with distributed activity. Merging models has been presented in our previous work [12] as an approach to address the problem. However, it was explained deeply intertwined within the context of a case study. This entanglement also occurs in other works, together with distinct vocabulary to describe the same concepts. It is clear that there are patterns in the intermediate phases of each approach, even if named differently. We address this abstraction by presenting a process to generalize rule-based models, such as decision trees or decision rules.

The remainder of this paper is structured as follows. Section 2 presents related work on generalizing rule-based models. Section 3 describes the process to generalize rule-based models and Sect. 4 provides a conclusion.

2   Related Work

It is reasonable to differentiate generalizing rule-based models from ensemble learning, which, at an initial glimpse, may appear similar. Ensemble learning [8] consists of using the predictions made by a number of base models to make a single prediction. In contrast, generalizing models consists of using a set of base models to create a single model, which is the only one making a prediction. The goals of each technique are also quite different. While in ensemble learning it is focused on improving accuracy, in generalizing models it is concerned in obtaining aggregated models without significantly affecting accuracy. Moreover, model interpretability is a goal per se for generalizing models but not for ensemble learning.

Approaches to generalizing models fall into two major categories: analytical and mathematical. Analytical approaches were first introduced by Williams [15] and consist of breaking down a set of models into rules and then assemble them in order to create a generalized model. On the other hand, mathematical approaches consist in applying a mathematical function to a group of models which results in the generalized model. The process described in this paper fits in the context of analytical approaches.

Analytical approaches emerged essentially from two contexts. The first was to create models for systems based on distributed environments, i.e., where the data sources were scattered across different locations. The problem was presented as "mining data that is distributed on distant machines, connected by low transparency connections" [2]. The second was a consequence of the growth in the amount of data collected by information systems. It became necessary to create models that could manage large datasets [7]. At the time there was a lack of available resources to handle the task, being described as "a very slow learning process sometimes overwhelming the system memory" [4] or "the emergence of datasets exceeding available memory" [1].

In problems with naturally distributed data, every location has its own local dataset with identical format and structure. These are moved over a channel to a centralized location where they are joined into a monolithic dataset, i.e., a non-distributed dataset stored in a single location. A generalized model is then created using all available data. Still, such scenario presents a major problem: moving data may be unsafe, expensive or simply impossible due to its large volume. An alternative of moving data is to move the models instead. Models are created in each location, then moved through a channel to a centralized location, where they are combined in a generalized model [2]. In problems with the need to create models from large datasets, it is essential to artificially create distributed data. This is achieved by breaking down a large dataset into as many individual datasets as necessary until it becomes possible to create a model for each [1]. Under such circumstances, all base models are combined into a generalized one.

Contrarily to analytical approaches, mathematical approaches are quite different from each other and were designed to solve specific problems. Kargupta and Park [6], motivated by the need to analyse and monitor time-critical data streams using mobile devices, proposed an approach to combine decision trees using the Fourier Transform. As the decision tree is a function, it can be represented in a frequency domain, resulting in the model *spectrum*. Models are combined by the adding their spectra. Gorbunov and Lyubetsky [3] combine models by constructing a *supertree*, the "nearest" on average to a given set of trees. The method is tested on the domain of analysis of the evolution trees of different species. In this context, the problem is to map a set of gene trees into a species tree (the average tree). Shannon and Banks [10] describe *Maximum Likelihood Estimate* (MLE) to combine a set of classification trees into a single tree by finding a *central tree*. The approach was applied to a set of classification trees obtained from biomedical data.

3   Generalization of Rule-Based Models

Generalization of rule-based models is presented as a sequential process with abstract parts and a few that can be specialized. Given a set of datasets $(D_1, \ldots, D_n)$, the corresponding base models $(M_1, \ldots, M_n)$ are trained and evaluated (using metric η_i). Then, all the base models are organized into groups $(G_1, \ldots, G_k)$ with a generalized model being created for each group $(\Omega_1, \ldots, \Omega_k)$. Finally the generalized models are evaluated using previously unused parts of the base models datasets (with metric σ_i). Figure 1 depicts a high-level view of the experimental set-up of the process, while Algorithm 1 presents it in more depth.

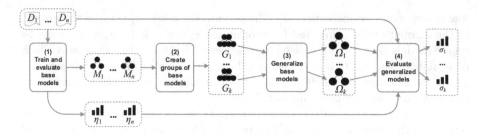

Fig. 1. Experimental set-up

A preliminary task of the process is the creation of 10 folds for each dataset ($\{f_i^1, ..., f_i^{10}\} \in D_i$). Each generalized model has to be evaluated using unseen data, i.e., data not used in the creation of base models. As each base model is to provide rules to a generalized model, one fold of its associated dataset is put aside destined to incorporate a test dataset to evaluate that same generalized model. As a consequence, to ensure it remains new data, this fold (denoted as λ) is never included in the data for creating or evaluating base models. Instead of choosing a specific fold, the process of generalizing models and subsequent evaluation is performed 10 times, each using a different fold. Each fold maps in an iteration (λ) of the evaluation cycle. Base models (M_i^λ) are created and then evaluated using the data in all folds except the λ fold ($D_i \setminus f_i^\lambda$).

Algorithm 1. Process to Generalize Rule-based Models

Input: Datasets = $\{D_1, \ldots, D_n\}$
Output: Improvement scores = $\{\sigma_1, \ldots, \sigma_k\}$
for λ such that $1 \leq \lambda \leq 10$ **do**
    for i such that $1 \leq i \leq n$ **do**
        $\{M_i^\lambda, \eta_i^\lambda\} \leftarrow$ TrainEvaluateBaseModel($D_i \setminus f_i^\lambda$)
    end for
    $\{G_1^\lambda, \ldots, G_k^\lambda\} \leftarrow$ CreateGroupsBaseModels($\{M_1^\lambda, \ldots, M_n^\lambda\}$)
    for j such that $1 \leq j \leq k$ **do**
        $\Omega_j^\lambda \leftarrow$ GeneralizeBaseModels(G_j^λ)
        $\sigma_j^\lambda \leftarrow$ EvaluateGeneralizedModel($\Omega_j^\lambda, \{f_1^\lambda, \ldots, f_p^\lambda\}, \{\eta_1^\lambda, \ldots, \eta_p^\lambda\}$)
    end for
end for
for j such that $1 \leq j \leq k$ **do**
    $\sigma_j \leftarrow \frac{1}{10} \sum_{\lambda=1}^{10} \sigma_j^\lambda$
end for

All base models are then organized into groups (G_j^λ), each to yield a generalized model (Ω_j^λ). Next, the evaluation test folds are assembled as test dataset for the generalized model ($\{f_1^\lambda, \ldots, f_p^\lambda\}$). The evaluation procedure takes the generalized model, the test dataset and the base models performances ($\{\eta_1^\lambda, \ldots, \eta_p^\lambda\}$

(p denoting the number of models in the group) resulting in an improvement score of the generalized model (σ_j^λ). This aim of this metric is to estimate whether there is gain (if positive) or loss (if negative) in predictive quality relative to the base models. Finally, as the evaluation cycle is replicated 10 times, the improvement scores of each generalized model are averaged across all iterations yielding the overall improvement score (denoted as σ_j).

3.1 Train and Evaluate Base Models

Models are rule-based classifiers, i.e., a set of IF-THEN rules. Due to one of the folds being reserved for evaluating the generalized model, base models are evaluated using 9-fold cross-validation set-up [11]. The evaluation score is conceptually denoted as η_i (which may be embodied, for example, with the F1-score [5]).

3.2 Create Groups of Base Models

In this procedure, the base models are gathered into groups. Models can be grouped reflecting a business driven criterion. For example, if a company is interested in knowing the performance of sales of its subsidiaries, it may want to group the models by geographic zone. Alternatively, there are applications where the creation of groups may be completely automated. An example is by criteria related to the complexity of the model, as the number of rules. In such cases, clustering techniques can be used to assist the creation of the groups. There may be applications where there is no need to create groups. Nevertheless, in order to maintain the process generic, it is considered that there is a single group with all the models.

3.3 Generalize Base Models

In this procedure, the base models in each group are generalized resulting in a new model, as described in Algorithm 2. Keeping up with the generality, it is required for the process to be independent of the language of the base models. In other words, it should be applied whether rules are extracted from decision trees or laid out in any other format. A possibility is for rules to be be represented as rows in a decision table with columns specifying the independent (x_i) and target ($\hat{y}$) variables. Therefore, before pursuing the combination of base models (M), these are converted to decision tables (denoted as T), and then generalized sequentially. Depending on the approach chosen to combine decision tables, there may be circumstances that generate an empty decision table. If so, the procedure skips that attempt and carries on selecting the next decision table to combine with the last one that succeeded (T_ω). After all the decision tables in the group are scanned, the final generalized decision table is converted back into the same language as the base models, yielding the generalized model (Ω). The next subsections detail these operations.

Algorithm 2. Generalize base models

Input: Group of base models $\{M_1, \ldots, M_p\}$
Output: Generalized model Ω
$T_\omega \leftarrow$ ExtractRules(M_1)
for i such that $2 \leq i \leq p$ **do**
    $T_\theta \leftarrow$ CombineRules(T_ω, ExtractRules(M_i))
    if $T_\theta \neq \emptyset$ **then**
        $T_\omega \leftarrow T_\theta$
    end if
end for
$\Omega \leftarrow$ BuildModel(T_ω)

Extract Rules. This operation extracts the underlying rules of a model as rows in a decision table (T), using an approach in accordance with its language.

Combine Rules. This operation attempts to combine the rules if a pair of decision tables (T_1 and T_2) into one (T_θ), with the steps presented in Algorithm 3.

Algorithm 3. Combine rules

Input: Decision tables T_1 and T_2
Output: Combined decision table T_θ
$T_\theta \leftarrow$ CreateRules(T_1, T_2)
if $T_\theta \neq \emptyset$ **then**
    $T_\theta \leftarrow$ JoinRules(ResolveConflicts(T_θ))
end if

The operation *Create rules* implies a specific approach to derive the rules of the combined table. A common example is the intersection of the inner product of the rules of both tables [1,2,12]. The operation is replicated until all rules from both tables are combined. A possible consequence is that none of the rules of both tables overlap, resulting in the intersection to be an empty set. If this occurs the process stops. Although the operation is illustrated with intersection, it is important to highlight that it is generic, i.e., it can be carried out with any another function. A conflict exists if a pair of overlapping rules of T_1 and T_2 do not agree on the target variable value. The operation *Resolve conflicts* selects, for each conflict found, which value should be set to the target variable of the new rule. For example, an approach is to assign the target value of the rule covering the larger volume in the multidimensional decision space [1]. Another is to select the one created with more examples [12]. After this operation, the resulting decision table has no conflicts. The operation *Join rules* attempts to decrease the number of rules by identifying adjacent rules in the multidimensional decision space sharing the same class in the target variable. These can be joined together, thus reducing the number of rules.

Build Model. This operation converts a decision table back to the base model representation. For example, if the base models are decision trees, then the generalized model should also be a decision tree. This task presents unexpected challenges. An inevitable consequence of repeatedly changing and removing decision rules along the combination process is a final decision table frequently failing to cover the entire multidimensional space. An approach consists in artificially generating examples falling into each decision region of the final generalized decision table T_ω [12]. The examples of all rules are gathered in a dataset D^{T_ω} from which a model is trained (Ω).

3.4 Evaluate Generalized Models

In this procedure, a generalized model is evaluated following the steps in Algorithm 4. The predictive quality of the generalized models is measured by an *improvement score* (denoted as σ).

Algorithm 4. Evaluate generalized model

Input: Gener. model = Ω, Test folds = $\{f_i, \ldots, f_p\}$, Perf. base models = $\{\eta_i, \ldots, \eta_p\}$
Output: Improvement score of generalized model = σ
for i such that $1 \le i \le p$ **do**
    $\Delta_i \leftarrow$ EvaluateModel$(\Omega, f_i) - \eta_i$
end for
$\sigma \leftarrow \frac{1}{p} \sum_{i=1}^{p} \Delta_i$

The fold that was put aside in each base model is used as test data to evaluate the generalized model. Evaluation consists in using the generalized model to make predictions on the test data and then comparing them with the true values of the target variable. The evaluation metric has to be the same as the one used to evaluate base models (e.g. if the F1-score was chosen to evaluate base models, then it should also be used to evaluate the generalized ones). As the aim is to estimate the variation in predictive quality of replacing the base models with a generalized one, the difference of performances (Δ_i) is assessed. If positive, then the generalized model performs better than the base model, otherwise, it performs worse. The cycle is replicated for all folds coming from each dataset of the base models associated with the generalized model. The overall performance of the generalized model results from the average of the differences of performances relative to all base models (σ) in the original group.

4 Conclusions

Generalizing rule-based models has emerged from approaches to solve different problems in particular contexts. Analytical approaches, which separate the rules of a set of models and then recombine them, although presented in a variety of forms, can be abstracted to a generic method. The main contribution of this paper is to describe a process that sequences the main procedures and then identifies the operations that can be deployed in different ways.

The representation of models as a set of decision rules facilitates the process of generalizing them. Then, the sub-problems of how to combine decision rules, resolve class conflicts of the target variable in overlapping rules, and build the generalized rule-based model remains open to different approaches, without loss of generality. Generalized models are evaluated by assessing their ability to replace the base models. Although the set-up to evaluate generalized models is part of the process, the evaluation metric itself is generic.

References

1. Andrzejak, A., Langner, F., Zabala, S.: Interpretable models from distributed data via merging of decision trees. In: Proceedings of the 2013 IEEE Symposium on Computational Intelligence and Data Mining. IEEE (2013)
2. Bursteinas, B., Long, J.: Merging distributed classifiers. In: Proceedings of the 5th World Multiconference on Systemics, Cybernetics and Informatics (2001)
3. Gorbunov, K., Lyubetsky, V.: The tree nearest on average to a given set of trees. Probl. Inf. Transm. **47**(3), 274–288 (2011)
4. Hall, L., Chawla, N., Bowyer, K.: Combining decision trees learned in parallel. In: Working Notes of the KDD-97 Workshop on Distributed Data Mining, pp. 10–15 (1998)
5. Han, J., Kamber, M., Pei, J.: Data Mining: Concepts and Techniques. Morgan Kaufmann, San Francisco (2011)
6. Kargupta, H., Park, B.: A fourier spectrum-based approach to represent decision trees for mining data streams in mobile environments. IEEE Trans. Knowl. Data Eng. **16**, 216–229 (2004)
7. Maimon, O., Rokach, L.: Data Mining and Knowledge Discovery Handbook, 2nd edn. Springer, Boston (2010). https://doi.org/10.1007/978-0-387-09823-4
8. Opitz, D., Maclin, R.: Popular ensemble methods: an empirical study. J. Artif. Intell. Res. **11**, 169–198 (1999)
9. Quinlan, J.: Induction of decision trees. Mach. Learn. **1**(1), 81–106 (1986)
10. Shannon, W.D., Banks, D.: Combining classification trees using MLE. Stat. Med. **18**(6), 727–740 (1999)
11. Stone, M.: Cross-validatory choice and assessment of statistical predictions. J. Roy. Stat. Soc.: Ser. B **36**(2), 111–147 (1974)
12. Strecht, P., Mendes-Moreira, J., Soares, C.: Merging decision trees: a case study in predicting student performance. In: Luo, X., Yu, J.X., Li, Z. (eds.) ADMA 2014. LNCS (LNAI), vol. 8933, pp. 535–548. Springer, Cham (2014). https://doi.org/10.1007/978-3-319-14717-8_42
13. Weiss, S., Indurkhya, N.: Optimized rule induction. IEEE Expert **8**(6), 61 (1993)
14. Weiss, S.M., Indurkhya, N.: Rule-based machine learning methods for functional prediction. J. Artif. Intell. Res. **3**, 383–403 (1995)
15. Williams, G.: Inducing and combining multiple decision trees. Ph.D. thesis, Australian National University (1990)

Introducing Noise in Decentralized Training of Neural Networks

Linara Adilova[1,2(✉)], Nathalie Paul[1], and Peter Schlicht[3]

[1] Fraunhofer IAIS, Sankt Augustin, Germany
{linara.adilova,nathalie.paul}@iais.fraunhofer.de
[2] Fraunhofer Center for Machine Learning, Sankt Augustin, Germany
[3] Volkswagen Group Research, Wolfsburg, Germany
peter.schlicht@volkswagen.de

Abstract. It has been shown that injecting noise into the neural network weights during the training process leads to a better generalization of the resulting model. Noise injection in the distributed setup is a straightforward technique and it represents a promising approach to improve the locally trained models. We investigate the effects of noise injection into the neural networks during a decentralized training process. We show both theoretically and empirically that noise injection has no positive effect in expectation on linear models, though. However for non-linear neural networks we empirically show that noise injection substantially improves model quality helping to reach a generalization ability of a local model close to the serial baseline.

1 Introduction

The idea of noisy optimization originates from the physical process of annealing [14], where noise helps to stabilize the state of the system through changing it by small random perturbations. Injecting noise in various ways is exploited both in convex problems [3] and non-convex ones, specifically, in the training of neural networks [1]. Research empirically shows that noise injection into the neural network objective optimization benefits the training process and improves generalization [7,19]. In our work we consider the particular case of noise injection into the model parameters, i.e., the neural network weights.

Optimization of the objective function for various tasks can also be performed in a decentralized manner, i.e., multiple learners train on distributed data sources and synchronize according to a chosen schedule [11,13,17]. Decentralized training has a strong motivation coming from a growing amount of devices, e.g., mobile phones, with possibly privacy sensitive data sources and the ability to perform local computations. However, models obtained via synchronizing results of local training often do not reach the performance of serial baseline trained on a centralized dataset.

L. Adilova and N. Paul—These authors contributed equally.

A. Monreale et al. (Eds.): ECML PKDD 2018 Workshops, CCIS 967, pp. 37–48, 2019.
https://doi.org/10.1007/978-3-030-14880-5_4

In order to tune the quality of the distributed training setup various synchronizing protocols were investigated, which address communication timing and aggregating operators [11–13]. We consider using an existing approach for improving serial training, namely noise injection, for the case of decentralized setup in order to reach higher accuracies of the local models.

We give an overview of the related research in Sect. 2. For prior theoretical investigation we prove in Sect. 3 that in the special case of linear neural networks zero-mean noise does not have an adversarial effect on the results of training, because it cancels out in expectation. This is supported by experiments in Sect. 4.1. Since non-linear neural networks have more practical usage, in Sect. 4.2 we empirically study noise injection to non-linear neural networks trained in the decentralized manner and show that it improves performance. The simplest case of noise injection to the neural network weights is initialization noise. Experiments of McMahan et al. [17] suggest that averaging of independently trained neural networks on different local datasets results in a model that performs worse than any model of the local learners. However, when averaging periodically, our experiments show that noisy initialization is indeed beneficial. Further experiments in Sect. 4.2 are concerned with noise injection during each synchronization step and show an improved quality of the trained models compared to the non-noisy setup for two considered classification tasks. Section 5 summarizes the research results and suggests possible future work.

2   Related Work

Various ways of noise injection into the process of neural networks training were thoroughly investigated in multiple research works. For example, Bishop [4] has shown that additive noise on the inputs is equivalent to a regularization term in a loss function if the noise amplitude is sufficiently small. An [1] explored noise injection to the inputs, outputs, and weights of neural networks. He observed that additive noise injection to the weights on each update step, either in on-line or in batch gradient descent optimization, leads to higher generalization of the learned solution. It was shown that noise added to the inputs affects only the smoothness of the resulting function, while noise injected to the weights also punishes large values and activations. Injected output noise changes the loss function only by a constant value and thus does not affect the quality of the learned function. Later Wang and Principe [22] demonstrated that noise added to the desired signal affects the variance of weights and leads to faster convergence. A different approach to noise injection to the training process is disturbing the gradient updates. It is investigated, for example, in works of An [1] or Neelakantan et al. [19]. Empirical results show that adding decaying noise to gradients helps achieving a global minimum, but does not affect the generalization quality of the resulting network. Yet reaching a global minimum might be even harmful for generalization, since such solutions tend to overfit on the training data.

Decentralized training of neural networks on distributed data sources with different ways of synchronization is investigated in works of McMahan et al.

[17], Jiang et al. [10], Smith et al. [20] and some others. The issue closest to our research is the initialization aspect considered by McMahan et al. [17]. The process of the neural network objective optimization is sensitive to the initial state of the model as it was presented by Kolen and Pollack [15]. According to McMahan et al. [17] the averaging is harmful if initialization of the locally trained models is different, which corresponds to injection of random noise to initially equal states. Nevertheless the effect of initialization noise together with periodic averaging was not investigated.

3   Noisy Averaging

To start our theoretical investigation we consider the simplest case of models, that is linear models, and analyze the effect of noise injection into their parameters for a common training setup. Here we present the summarization of used basic concepts and a proof that noise injection in the considered special case for linear models has no detrimental effect on the final model.

3.1   Basic Concepts

The task of training a neural network is an optimization problem which is commonly solved by using *stochastic gradient descent* (SGD) or SGD-based algorithms. SGD is an on-line optimization algorithm in which a local minimum of the objective function is determined by moving in the negative direction of its gradient [6]. SGD is derived from the gradient descent algorithm and estimates the gradient by considering only one training example. Let X denote the *input space*, Y the *output space*, $\mathcal{F}$ the *model parameters space* and $f \in \mathcal{F}$ the model parameters. The SGD update rule for minimizing an *objective function* $\ell : \mathcal{F} \times X \times Y \to \mathbb{R}_+$ reads

$$\varphi_\eta^{SGD}(f) = f - \eta \nabla \ell(f, x, y),$$

where the learning rate η controls the step size.

In our work we consider *mini-batch SGD* which approximates the gradient by taking batch size B many training examples into account:

$$\varphi_{B,\eta}^{mSGD}(f) = f - \eta \sum_{j=1}^{B} \nabla \ell^j(f), \tag{1}$$

where $\ell^j(f) = \ell(f, x_j, y_j)$.

An example of a loss function is *squared loss* $\ell^j(f) = 1/2\left(\langle f, x_j \rangle - y_j\right)^2$.

Choosing this loss function and computing its gradient, we can write the learning algorithm update rule in the following form:

$$\varphi_{B,\eta}^{mSGD}(f) = f - \eta \sum_{j=1}^{B} \left(\left(\langle f, x_j \rangle - y_j\right) x_j\right).$$

3.2   Periodic Averaging Protocol with Noise Injection

As a protocol for the theoretical investigation of the effect of noise injection in the decentralized setup we consider the periodic averaging protocol with zero-mean noise injection (Algorithm 1). Let Ψ denote a probability distribution with mean zero and $\epsilon_t \geq 0$ the time-dependent noise level factor.

Algorithm 1. Periodic averaging protocol with zero-mean noise injection

---

Input: batch size b

Initialization:

    local models $f_1^1, \ldots, f_1^m \leftarrow$ one random f

At round t:

  At learner i:

    observe $\left(x_{t,1}^i, y_{t,1}^i\right), \ldots, \left(x_{t,B}^i, y_{t,B}^i\right)$

    $\psi_i \leftarrow \Psi$

    update f_t^i using the learning algorithm $\varphi : f_{t+1}^i \leftarrow \varphi\left(f_t^i + \epsilon_t \psi_i\right)$

  If $t \bmod b = 0$:

    synchronize local models: $f_{t+1}^i \leftarrow \frac{1}{m} \sum_{i=1}^m f_{t+1}^i$

In the following we analyze the influence of noise injection on the behavior of the learning algorithm for the special case of linear models. For this case we can prove that adding zero-mean noise to the parameters optimized by mini-batch SGD with squared loss does not change the model parameters in expectation.

Lemma 1. *For a linear model let f_t denote the model parameters attained by using mini-batch SGD with squared loss and scaled additive zero-mean weight noise injection, i.e.*

$$f_{t+1} = \varphi_{B,\eta}^{mSGD}(f_t + \epsilon_t \psi).$$

Let g_t denote the model parameters attained by using common mini-batch SGD, i.e.

$$g_{t+1} = \varphi_{B,\eta}^{mSGD}(g_t).$$

If the learning algorithms are identically initialized, it holds

$$\mathbb{E}\left[f_t\right] = g_t \qquad \text{for all } t = 1, \ldots, T. \tag{2}$$

Proof. We use induction over t. The case $t = 1$ follows immediately since by initialization the model parameters trained with and without noise are the same. Applying the definitions above, the expected model parameters for mini-batch SGD with noise injection read

$$\mathbb{E}\left[f_t\right] = \mathbb{E}\left[\varphi_{B,\eta}^{mSGD}\left(f_{t-1} + \epsilon_t\psi\right)\right]$$

$$= \mathbb{E}\left[f_{t-1} + \epsilon_t\psi - \eta\sum_{j=1}^{B}\left(\left(\langle f_{t-1} + \epsilon_t\psi, x_{(t-1,j)}\rangle - y_{(t-1,j)}\right)x_{(t-1,j)}\right)\right]$$

$$= \epsilon_t\underbrace{\mathbb{E}\left[\psi\right]}_{=0} + \mathbb{E}\left[f_{t-1} - \eta\sum_{j=1}^{B}\left(\left(\langle f_{t-1}, x_{(t-1,j)}\rangle - y_{(t-1,j)}\right)x_{(t-1,j)}\right)\right]$$

$$- \underbrace{\mathbb{E}\left[\eta\sum_{j=1}^{B}\left(\epsilon_t\langle\psi, x_{(t-1,j)}\rangle x_{(t-1,j)}\right)\right]}_{=0},$$

where we used that ψ is centered. By employing the induction assumption we get $\mathbb{E}\left[f_{t-1}\right] = g_{t-1}$ and conclude

$$\mathbb{E}\left[f_t\right] = g_{t-1} - \eta\sum_{j=1}^{B}\left(\left(\langle g_{t-1}, x_{(t-1,j)}\rangle - y_{(t-1,j)}\right)x_{(t-1,j)}\right)$$

$$= \varphi_{B,\eta}^{mSGD}(g_{t-1})$$

$$= g_t.$$

$\square$

The equivalence of Algorithm 1 to the non-noisy periodic averaging protocol in expectation directly follows from the linearity of the expected value.

Corollary 1. *For linear models trained with mini-batch SGD and squared loss given identical initialization, the expected model obtained by the periodic averaging protocol with zero-mean noise injection is equivalent to the model obtained by the non-noisy periodic averaging protocol.*

For the specific considered case of linear models we observe that noise injection has no influence on the model parameters in expectation both for serial and distributed learning. This theoretical result is supported by empirical evidence in the following Section. In contrast, for non-linear models the research of An [1], Edwards and Murray [7] has shown that noise injection in serial training helps improving generalization. We conjecture that in the distributed training setup injecting noise into non-linear models might also improve generalization properties of the obtained solution compared to the non-noisy training. We leave a theoretical investigation of the effect of noise injection into non-linear models in distributed training setup for future work. To substantiate our conjecture, in the following Section we perform an empirical analysis of zero-mean noise injection in a distributed setup for non-linear neural networks.

4   Empirical Evaluation

To investigate the effect of noise injection into the neural networks trained in a decentralized manner we performed a set of experiments described further.

The decentralized setup of these experiments is periodic synchronization via averaging (cf. Algorithm 1). Apart from the distributed synchronized models two baselines are trained: a local model without any synchronization with other local learners (no-sync) and a model with full data centralization (serial). These baselines are necessary to assess the performance of the synchronizing local learners, since synchronization aims to reach the performance of the serial model and outperform the no-sync baseline. Noise injection into the serial baseline is also a subject of interest, allowing to compare the possible gains in generalization ability in centralized and distributed case. For evaluation of the synchronizing learners the last averaged model obtained during the training process is used. For evaluation of the no-sync baseline we pick one random model among locally trained ones.

Since we explore the effect of noise injection, we are interested in the behavior of models trained throughout several experiments that differ only by the used noise. Thus all the setups were run 10 times without using fixed random seed. This produces an indication of how distribution of possible outcomes of the training process looks like. The results are presented in the form of box plots. Here the box shows the observed values from the first to the third quartile, with a line at the median. The whiskers show the range of the results and points are representing outliers.

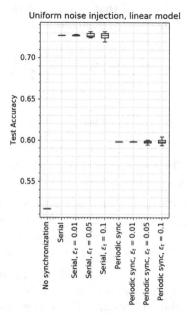

Fig. 1. Effect on test accuracy of noise injection to local learners and to the serial baseline.

4.1 Linear Neural Networks

First set of experiments is performed to empirically evaluate the theoretical result obtained in Sect. 3. We employed a linear neural network with three layers having 2, 64 and 1 neuron correspondingly for approximating the target column of SUSY dataset [2]. For linear model experiments the dataset was normalized thus having -1 and 1 as targets and accuracy was calculated with 0 threshold. The optimal training parameters determined were mini-batch of $B = 10$ examples, learning rate of $\eta = 1e{-}5$ for the serial and no-sync baselines and $\eta = 2.5e{-}5$ for the local learners. We employed squared loss as explained in Sect. 3. During the training each local learner was presented 20000 examples, while serial baseline had $m \cdot 20000$ examples with $m = 10$. The noise used for this experiment is additive uniform noise in the range $[-0.5, 0.5]$ with decay factor equal to the synchronization round number. Absence of decay factor was leading to fast overflow thus making experiments not runnable.

Figure 1 shows the test accuracy evaluated for 10 runs for the baselines and synchronizing learners on the independent test set of 1000000 examples. We can observe that larger noise injected into the weights of the models leads to larger variance of obtained accuracy at the same moment leaving the median value throughout different setups the same. This empirically supports the effect of noise cancellation in expectation for linear models in this training setup.

4.2 Non-linear Neural Networks

In the following we evaluate noise injection to the decentralized training of non-linear neural networks on the basis of two classification tasks. For our experiments we choose to add uniform noise in range $[-0.5, 0.5]$ and Gaussian noise in the same range.

Binary Classification. For preliminary evaluation of the approach for non-linear case we have considered the classification task on the SUSY dataset. In contrast to the linear case, the employed model architecture is a three-layered dense network with sigmoid activations. The first layer has 32 neurons, the second 64 and the output layer has 2 neurons with softmax activation. We have determined the optimal parameters of the non-noisy learning algorithm on a small fraction of the dataset. That is training mini-batch of $B = 10$ examples, learning rate of $\eta = 0.1$ for the serial and no-sync baselines and $\eta = 0.25$ for the local learners.

Initialization Noise. The simplest way of noise injection to the distributed neural networks is one step noise injection right after initialization. In McMahan et al. [17] it was shown that such noise together with one-time synchronization after local training results in a worse model than each local one in terms of training loss. We want to investigate whether periodic synchronization is capable of being more robust to initial noise injection.

With regard to Algorithm 1, initialization noise corresponds to choosing $\epsilon_1 > 0$ and $\epsilon_t = 0$ for all $t > 1$. It means that we add randomly sampled noise to each initial weight before starting the training process.

Figure 2 shows that both Gaussian and uniform initialization noise improves the serial and periodically synchronizing models in terms of cumulative training loss when using it up to some small extent ($\epsilon_t < 1$). On the contrary, higher levels of noise (e.g. $\epsilon_t = 5$) make training harder in each setup. Interestingly, a noise level of $\epsilon_t = 2$ for both Gaussian and uniform noise leads to a higher cumulative training loss in the serial model while the distributed setup benefits from it.

Even though the Gaussian distribution is a very popular choice for initializing neural networks (Glorot and Bengio [9]), adding large levels noise drawn from the normal distribution deteriorates the training process worse than uniformly distributed noise. One possible explanation is that the initial weights are already distributed normally according to the best practices and additional Gaussian noise intervenes with it in a destructive way.

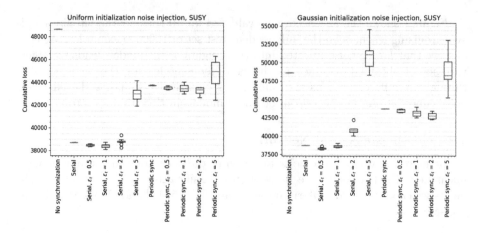

Fig. 2. Effect on cumulative training loss of uniformly and normally distributed noise added to the equal initializations of local learners and to the serial baseline.

Initialization noise experiments reveal that one-time noise injection to the initially equal model weights helps the training process since the cumulative training loss is decreasing. This motivates follow-up experiments with continuous noise injection.

Continuous Noise. Extending the initialization noise setup we now perform additional noise injection steps: Zero-mean noise gets injected to the local models' weights after every synchronization step. Formally, in Algorithm 1 this corresponds to setting $\epsilon_1 > 0$, and $\epsilon_{t+1} > 0$ for all t mod $b = 0$ and $\epsilon_t = 0$ otherwise. Following the work of Murray and Edwards [18] the noise is decaying and the decaying factor is equal to the index number of the synchronization step, i.e. noise level is given by ϵ_t/t.

We want to explore whether continuous noise injection improves the generalization ability of the resulting models in the distributed setup. Therefore we calculated the evaluation accuracy on an independent test set of 1000000 examples for each of the trained models. During training, each of the local learners i is presented 1000 examples from the training dataset, while the serial model sees $m \cdot 1000$ examples. Various setups together with evaluated validation accuracies are depicted as box plots in Fig. 3.

The evaluation shows that noise injection can substantially improve the generalization ability of the models. When comparing results of the setup with 10 and 20 learners, we see that a larger amount of learners leads to a larger spread of the results of the training process that can mean either a better generalization or to the contrary convergence to a worse model.

One can also observe that with growth of the uniform noise level the resulting test accuracy becomes more unstable, i.e., for having a possibly higher median we get a larger range of values. In the experiments, Gaussian noise is more stable than uniform noise while on the other hand it has more outliers below

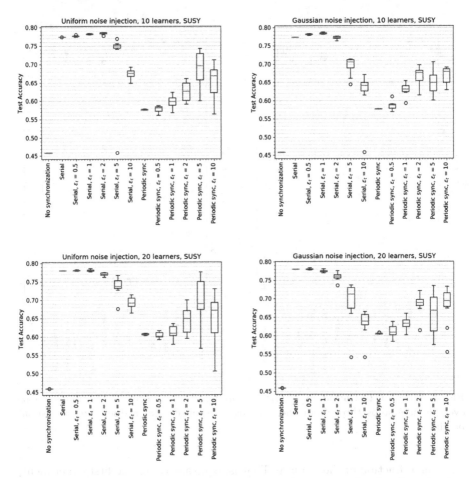

Fig. 3. Effect on test accuracy of injecting uniformly or normally distributed noise every b time steps throughout the training process of local learners and the serial baseline.

the median. The spread of the serial baseline with noise injection is very small compared to the distributed models. It might be interesting to investigate the reasons why noise has a more pronounced effect in the distributed setup than in the serial one.

Image Classification. To further investigate the effect of noise injection into non-linear models in a distributed setup we have chosen the classification task on the MNIST dataset [16]. The model architecture is more complicated than in the previous experiment. It has two dense layers with 512 neurons each and a dropout layer after each of them. The output layer performs a softmax activation to predict one of the ten classes. The activation of the dense layers is ReLU. We have determined the optimal parameters of the non-noisy learning algorithm

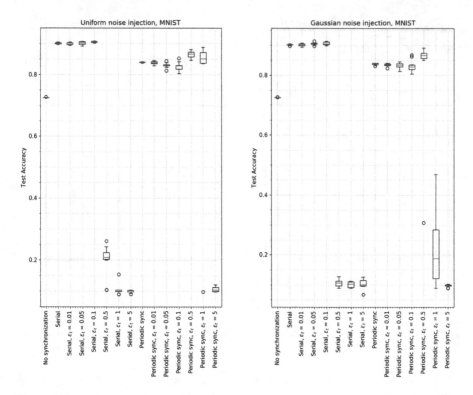

Fig. 4. Effect on test accuracy of injecting uniformly and normally distributed noise every b time steps throughout the training process of local learners and the serial baseline.

on a small fraction of the dataset. That is, equivalently to the first experiment, training mini-batch of $B = 10$ examples, learning rate of $\eta = 0.1$ for the serial and no-sync baselines and $\eta = 0.25$ for the local learners. During training each learner is presented 500 examples from the training set and evaluation is performed on the 10000 images of the test set.

We observe that in this experiment, compared to the previous, noise injection requires different level values in order to obtain the improved quality of the models (Fig. 4). More precisely, for noise levels $\epsilon_t \geq 1$ we get prohibitively low test accuracies for both uniform and Gaussian noise. One possible explanation is that the ReLU activation is much more sensitive to noise disturbance of the weights compared to the bounded sigmoid function which we have used in the first experiment. Moreover the dropout layers might be contributing to this effect since dropout is also supposed to prevent a model from overfitting [21]. More refined research on this effect is left for future work. Concentrating on lower levels of noise ($\epsilon_t < 0.5$), we see that we can again improve the generalization ability of the trained models. In the distributed setup this effect is more pronounced than in the serial one similarly to the first set of experiments.

5   Conclusion

The research presented in this paper investigates noise injection into neural networks, trained in a decentralized manner on distributed data sources. We have proven that for linear models in a common training setup zero-mean noise injection retains the results of the non-noisy setup. We have performed experiments to empirically underline this theoretical statement. Further experiments show that with non-linear models in a distributed setup noise injection improves the quality of the models. The evaluation shows that indeed carefully chosen levels of noise have a positive effect on the generalizing abilities of the synchronized models. It might be explained by the fact that noise enforces wider exploration of the space of solutions [5,8] which is an interesting subject for further investigation. Also, experiments show that the impact of noise in the distributed training is even greater than in the serial case.

Future research could investigate the theoretical background of noise injection to non-linear models in a decentralized training setup as well as the effect of various network architectures and training parameters. A promising framework for studying noise injection effects is regularization theory, since injected noise can be described as a regularization term that is added to the loss function [1,4].

References

1. An, G.: The effects of adding noise during backpropagation training on a generalization performance. Neural Comput. **8**(3), 643–674 (1996)
2. Baldi, P., Sadowski, P., Whiteson, D.: Searching for exotic particles in high-energy physics with deep learning. Nat. Commun. **5**, 4308 (2014)
3. Ben-Tal, A., Nemirovski, A.: Robust convex optimization. Math. Oper. Res. **23**(4), 769–805 (1998)
4. Bishop, C.M.: Training with noise is equivalent to Tikhonov regularization. Neural Comput. **7**(1), 108–116 (1995)
5. Chen, B., Deng, W., Du, J.: Noisy softmax: improving the generalization ability of DCNN via postponing the early softmax saturation. In: The IEEE Conference on Computer Vision and Pattern Recognition (CVPR) (2017)
6. Chung, K.L., et al.: On a stochastic approximation method. Ann. Math. Stat. **25**(3), 463–483 (1954)
7. Edwards, P.J., Murray, A.F.: Fault-tolerance via weight-noise in analogue VLSI implementations—a case study with Epsilon. IEEE Proc. Circ. Syst. II: Analog Digit. Sig. Process. **45**(9), 1255–1262 (1998)
8. Gan, Z., Li, C., Chen, C., Pu, Y., Su, Q., Carin, L.: Scalable Bayesian learning of recurrent neural networks for language modeling. In: Proceedings of the 55th Annual Meeting of the Association for Computational Linguistics: Long Papers, vol. 1, pp. 321–331 (2017)
9. Glorot, X., Bengio, Y.: Understanding the difficulty of training deep feedforward neural networks. In: Proceedings of the Thirteenth International Conference on Artificial Intelligence and Statistics, pp. 249–256 (2010)
10. Jiang, Z., Balu, A., Hegde, C., Sarkar, S.: Collaborative deep learning in fixed topology networks. In: Advances in Neural Information Processing Systems, pp. 5904–5914 (2017)

11. Kamp, M., Boley, M., Keren, D., Schuster, A., Sharfman, I.: Communication-efficient distributed online prediction by dynamic model synchronization. In: Calders, T., Esposito, F., Hüllermeier, E., Meo, R. (eds.) ECML PKDD 2014. LNCS (LNAI), vol. 8724, pp. 623–639. Springer, Heidelberg (2014). https://doi.org/10.1007/978-3-662-44848-9_40

12. Kamp, M., Boley, M., Missura, O., Gärtner, T.: Effective parallelisation for machine learning. In: Advances in Neural Information Processing Systems, pp. 6477–6488 (2017)

13. Kamp, M., Bothe, S., Boley, M., Mock, M.: Communication-efficient distributed online learning with kernels. In: Frasconi, P., Landwehr, N., Manco, G., Vreeken, J. (eds.) ECML PKDD 2016. LNCS (LNAI), vol. 9852, pp. 805–819. Springer, Cham (2016). https://doi.org/10.1007/978-3-319-46227-1_50

14. Kirkpatrick, S., Gelatt, C.D., Vecchi, M.P.: Optimization by simulated annealing. Science 220(4598), 671–680 (1983)

15. Kolen, J.F., Pollack, J.B.: Back propagation is sensitive to initial conditions. In: Advances in Neural Information Processing Systems, pp. 860–867 (1991)

16. LeCun, Y.: The MNIST database of handwritten digits (1998). http://yann.lecun.com/exdb/mnist/

17. McMahan, H.B., Moore, E., Ramage, D., Hampson, S., et al.: Communication-efficient learning of deep networks from decentralized data. arXiv preprint arXiv:1602.05629 (2016)

18. Murray, A.F., Edwards, P.J.: Enhanced MLP performance and fault tolerance resulting from synaptic weight noise during training. IEEE Trans. Neural Netw. 5(5), 792–802 (1994)

19. Neelakantan, A., et al.: Adding gradient noise improves learning for very deep networks. arXiv preprint arXiv:1511.06807 (2015)

20. Smith, V., Chiang, C.K., Sanjabi, M., Talwalkar, A.S.: Federated multi-task learning. In: Advances in Neural Information Processing Systems, pp. 4424–4434 (2017)

21. Srivastava, N., Hinton, G., Krizhevsky, A., Sutskever, I., Salakhutdinov, R.: Dropout: a simple way to prevent neural networks from overfitting. J. Mach. Learn. Res. 15(1), 1929–1958 (2014)

22. Wang, C., Principe, J.C.: Training neural networks with additive noise in the desired signal. IEEE Trans. Neural Netw. 10(6), 1511–1517 (1999)

3rd Workshop on IoT Large Scale Machine Learning from Data Streams

Preface

Workshop Description

The volume of data is rapidly increasing due to the development of the technology of information and communication. This data comes mostly in the form of streams. Learning from this ever-growing amount of data requires flexible learning models that self-adapt over time. Traditional one shot memory based learning methods trained offline from a static historic data cannot cope with evolving data streams. This is because firstly, it is not feasible to store all incoming data over time and secondly the generated models become quickly obsolete due to data distribution changes, also known as 'concept drift'. The basic assumption of offline learning is that data is generated by a stationary process and the learning models are consistent with future data. However, in multiple applications like web mining, social networks, network monitoring, sensor networks, telecommunications, financial forecasting etc., data samples arrive continuously as unlimited streams often at high-speed. Moreover, the phenomena generating these data streams may evolve over time. In this case, the environment in which the system or the phenomenon generated the data is considered to be dynamic, evolving or non-stationary.

Learning methods used to learn from data generated by dynamically evolving and potentially non-stationary processes must take into account many constraints: (pseudo) real-time processing, high-velocity, and dynamic multi-form change such as concept drift and novelty. In addition in data streams scenarios, the number of classes is often unknown in advance. Therefore, new classes can appear any time and they must be detected and the predictor structure must be updated.

Therefore, data generated by phenomena in dynamical environments are characterized by: (1) potentially unlimited size, (2) sequential access to data samples in the sense that once an observation has been processed, it cannot be retrieved easily unless it is explicitly stored in memory and (3) unpredictable, dependent and not identical distributed observations.

Consequently, learning from streams of evolving and unbounded data requires developing new algorithms and methods able to learn under the following constraints: (1) random access to observations is not feasible or it has high costs, (2) memory is small with respect to the size of data, (3) data distribution or phenomena generating the data may evolve over time, which is known as concept drift and (4) the number of classes may evolve overtime.

It is worthwhile to emphasize that streams are very often generated by distributed sources, especially with the advent of Internet of Things and therefore processing them centrally may not be efficient especially if the infrastructure is large and complex. Scalable and decentralized learning algorithms are potentially more suitable and efficient.

This combined tutorial and workshop aimed at discussing the problem of learning from data streams generated by evolving non-stationary processes. It overviewed the advances of techniques, methods and tools that are dedicated to manage, exploit and interpret data streams in non-stationary environments. In particular, the event examined the problems of modeling, prediction, and classification based on learning from data streams.

The workshop aimed at presenting new research advances related to data streams processing. The complementarity between these methods and tools, together with methods and techniques discussed during the tutorial, was investigated in order to define new suggestions to develop and improve these methods as well as defining new application domains.

December 2018   Joao Gama

Organization

Workshop Chairs

Moamar Sayed-Mouchaweh   Computer Science and Automatic Control Labs,
     High Engineering School of Mines, Douai, France
Albert Bifet   LTCI, Telecom ParisTech; Paris, France
João Gama   Laboratory of Artificial Intelligence and Decision
     Support, University of Porto, Porto, Portugal
Rita Ribeiro   Laboratory of Artificial Intelligence and Decision
     Support, University of Porto, Porto, Portugal

Program Committee Members

Edwin Lughofer   Johannes Kepler University of Linz, Austria
Sylvie Charbonnier   Université Joseph Fourier-Grenoble, France
Bruno Sielly Jales Costa   IFRN, Natal, Brazil
Fernando Gomide   University of Campinas, Brazil
José A. Iglesias   Universidad Carlos III de Madrid, Spain
Anthony Fleury   Mines-Douai, Institut Mines-Télécom, France
Teng Teck Hou   Nanyang Technological University, Singapore
Plamen Angelov   Lancaster University, UK
Igor Skrjanc   University of Ljubljana, Slovenia
Indre Zliobaite   Aalto University, Finland
Elaine Faria   Univ. Uberlandia, Brazil
Mykola Pechenizkiy   TU Eindonvhen, Netherlands
Raquel Sebastião   Univ. Aveiro, Portugal

Keynote Speaker

Bernhard Pfahringer   University of Waikato

Program Committee

Thiago Andrade   INESC TEC
Albert Bifet   LTCI, Telecom ParisTech
Brais Cancela   Universidade da Coruña
Douglas Castilho   Federal University of Minas Gerais
Jonathan De Andrade Silva   University of Sao Paulo at Sao Carlos

Query Log Analysis: Detecting Anomalies in DNS Traffic at a TLD Resolver

Pieter Robberechts[1]([⊠]), Maarten Bosteels[2], Jesse Davis[1], and Wannes Meert[1]

[1] Department of Computer Science, KU Leuven, Leuven, Belgium
`pieter.robberechts@cs.kuleuven.be`
[2] DNS Belgium vzw, Leuven, Belgium

Abstract. We present QLAD, an anomaly detection system that is designed for the high query volume and the specific nature of DNS traffic at a TLD resolver. QLAD integrates three components that implement the complete anomaly detection process, ranging from the ingression of raw traffic data to the visualisation of detected anomalies. With an initial analysis of query logs from the Belgian ccTLD registry, we showed that QLAD can archive data compactly, has a low computational cost and can detect a wide range of anomalies. We found several anomalies that are of interest to the registry operator, such as domain enumerations and DoS attacks. Other anomalies were caused by benign applications with unique traffic patterns. A user interface helps to distinguish these, but correctly identifying all anomalies remains a difficult and tedious task.

Keywords: Anomaly detection · DNS · Internet security

1 Introduction

The Domain Name System (DNS) is an Internet service that translates domain names into IP addresses [19]. While the Internet's infrastructure is based on IP addresses, domain names are alphanumerical to make them easier to remember. Every time you use a domain name, a DNS service must translate the name to the corresponding IP address. For example, the domain name www.example.com might translate to 198.105.232.4.

At its core, DNS is a distributed and hierarchical database. Each level corresponds to a part in the domain name separated by dots and uses so-called authoritative name servers to provide pointers to the next lower level. At the top of the DNS hierarchy are 13 root name servers, which contain pointers to the name servers for all the generic top-level domains (gTLD) such as .com and .org as well as country-specific top-level domains (ccTLD) such as .uk, .be or .fr. Each TLD is managed by a registry operator which is responsible for the registration of new domain names and the resolution of queries for these domains. The other levels in this hierarchy follow the same principle.

On one hand, these TLD registries are themselves an interesting target for attackers. Given that most of the communication within the Internet starts with

© Springer Nature Switzerland AG 2019
A. Monreale et al. (Eds.): ECML PKDD 2018 Workshops, CCIS 967, pp. 55–67, 2019.
https://doi.org/10.1007/978-3-030-14880-5_5

several DNS lookups, a successful attack on servers high up in the DNS hierarchy can have far-reaching consequences. This was illustrated by the Dyn cyberattack that took place on October 21, 2016. It involved multiple distributed denial-of-service (DDoS) attacks targeting systems operated by DNS provider Dyn, which caused major Internet platforms and services to be unavailable to large swathes of users in Europe and North America [20]. On the other hand, TLD registries can leverage their unique position high up in the DNS hierarchy to detect threats against other stakeholders within their zone. Many threats such as phishing campaigns, spam attacks and Command and Control (C&C) communication used by botnets can be observed in their DNS traffic.

The TLD registries' business model relies on a high availability and resilience of their DNS service, as well as the high reputation and trustworthiness of their domain. Thus, it is in the registries' interest to quickly identify attacks targeting or abusing their infrastructure, and to identify misuse of their domain names. Nevertheless, threat detection is still mainly a post-hoc task. Most malicious behaviour is only observed after it has a noticeable impact on the service.

Using the Belgian TLD registry (DNS Belgium) as a use case, we will illustrate how proactive and real-time analysis of the continuous stream of DNS query logs can contribute to the resilience and security of a TLD registry's service. This paper will discuss the design and implementation of a query log analysis platform called *QLAD* (Query Log Anomaly Detection) that is able to detect attacks and other suspicious behaviour at a TLD resolver in near real time. *QLAD* performs a first set of attack and anomaly detection functions and it offers an interface for reporting and inspecting the detected anomalies. All code is open source and available at https://github.com/DNSBelgium/qlad.

To summarize, this paper makes the following contributions: (i) Highlight a number of data science challenges that we encountered while solving this problem; (ii) Discuss two statistical anomaly detection algorithms that jointly detect a wide range of threats in DNS query logs; (iii) Propose an architecture for storage, analysis and presentation of DNS query logs; (iv) Validate the approach on a real-world data set.

2   Related Work

SIDN Labs, the research unit of the Netherlands' TLD registry, were the first to present a complete framework that enables registries to increase the security and stability of their TLD [11]. However, compared to the approach presented in this work, they focus on the detection of domain names used in malicious activities instead of the detection of attacks against the DNS infrastructure. The operator of the .uk TLD developed Turing [21], a system that appears to be similar. However, Turing is a commercial solution and there is little publicly available information about its functionality and technical implementation.

Furthermore, there is scattered prior work on individual components for data storage, anomaly detection and thread mitigation. Traffic to ccTLD name servers produces gigabytes of traffic data each day. On the one hand, researchers have resorted to Hadoop-based solutions [3,4,16,32] as a way to address the high volume of data. Such cluster solutions are designed for long-term data storage and to support interactive queries on the stored data. On the other hand, tools such as the DNS Statistics Collector (DSC) [9] aggregate the data before archiving, reducing it to a size which can be stored in a traditional database.

Existing research about detecting the discussed attack vectors in the upper levels of the DNS hierarchy is limited. Most research focuses on anomaly detection in recursive resolvers and small authoritative resolvers [1,23,30]. Mostly, these methods do not scale to the level of a ccTLD. A couple of simple ideas are applicable at the TLD level, however. These methods are all based on the detection of changes in the frequency of packets with certain attributes [12,13,33]. For example, Karasaridis et al. [12] uses cross entropy to detect significant changes in the distribution of packet sizes. Considerably more research has been done on the broader scope of network anomaly detection [6]. These methods search for anomalies in IP or TCP traffic, but some can be applied on DNS traffic too [8].

3   Detecting Anomalies in DNS Server Logs

The core problem which we aim to solve is anomaly detection in a continuous stream of DNS query logs. We claim that due to the variety of types of anomalies this problem cannot be solved with a single approach. On the one hand, we need to analyse individual traffic flows to enable the detection of low-volume anomalies in the continuous stream of valid traffic. On the other hand, we need to look at global traffic patterns to detect anomalies that span multiple flows.

In this section, we firstly discuss the challenges posed by the task at hand and secondly introduce two complementary anomaly detection algorithms by which we address these challenges, referred to as respectively *QLAD-flow* and *QLAD-global*. Finally, we describe how to integrate these anomaly detection algorithms in a complete anomaly detection framework.

3.1   Challenges

Analysing DNS server logs to identify anomalies poses a number of significant and non-trivial data science challenges.

Volume of data. Each server has to process about 1,000 queries per second. This has two important consequences. First, efficiency and scalability are crucial: any algorithms used for analysis should have a low computational cost and should scale well to increasing traffic volumes. Secondly, due to the high volume of valid traffic, it is easy to mask malicious traffic.

No labelled data. A second challenge is the lack of labelled or clean training data. We have only access to the raw server logs. It is practically infeasible to manually label all anomalies in these logs. They are hidden between thousands of valid queries and one anomaly may correspond to thousands of unique queries, each valid on their own (*e.g.*, a DoS attack with spoofed source IP addresses). Consequently, we can not use any supervised machine learning algorithms to learn a model for anomalous or valid traffic patterns. Moreover, any other algorithm should be robust enough to deal with a certain amount of unidentified anomalies in the training data. Also, evaluating anomaly detectors is difficult, since we can not compare our results to a ground truth.

Adversarial setting. There is a wide range of possible attack vectors, which are under constant evolution. Attackers have often successfully modified their attack vectors to circumvent new security patches [28] and extensions to the DNS protocol have introduced new vulnerabilities [2]. A straightforward and popular approach to solve the problem is the signature-based technique [17], which defines a separate model (signature) for each popular attack. This technique requires prior knowledge and new types of attacks can go undetected [17]. Moreover, generating a good signature for an anomaly is difficult [10]. The signatures should be general enough to capture small variations on the targeted attack vector, while being specific enough to allow valid traffic. Such models are often easy to mislead with a small change to the attack method [7].

Nature of DNS traffic. DNS traffic is subject to both periodic and slow varying trends in terms of the volume, type, origin and content of queries. The number of user generated DNS queries decreases at night, over the weekend and during holidays. Administrative queries like SOA type queries are less affected by these factors. Moreover, apart from trends in the global amount of traffic, individual domains often have unique traffic patterns. Furthermore, the unique characteristics of the DNS data flow, such as small in packet size and little in message amount, make it more difficult to distinguish anomalous behaviours from normal ones [33].

Packet spoofing. An attacker can easily replace some fields in the DNS queries by random values. This will make it more difficult to identify anomalies, since all queries will seem unrelated.

3.2 QLAD-flow

The first algorithm's goal is to identify low volume anomalies. Therefore, it splits the traffic into different flows, for example based on the source IP of the DNS query. The partitioning makes it possible to detect anomalies which are otherwise hidden in the global flow of traffic. We employ the statistical anomaly detection approach proposed by Dewaele et al. [8]. We first briefly describe this method for completeness, and next highlight precisely what is different.

The algorithm analyses a continuous stream of packets within a sliding time-window of duration T. Each packet is identified by its time of arrival and a set of packet attributes (*i.e.*, source IP, destination IP, source port and destination port). First, all packets within a window are hashed to N buckets, using the packet attributes as hashing keys. Second, for each bucket, the algorithm counts the number of packets that arrive during a short aggregation period. This is done for multiple aggregation levels, such that each bucket is transformed to multiple series of packet counts. Each of these series is modelled by Gamma distributions $\Gamma_{\alpha,\beta}$. Then, we can estimate the average value and variance for the α and β parameters for each aggregation level and identify the buckets for which the α and β values deviate more than a given threshold. Finally, the algorithm repeats this procedure with different universal hash functions. Each hash function will result in a different set of anomalous buckets. The intersection of all these buckets should correspond to the set of all anomalous packets.

The method was originally designed to detect low-intensity anomalies in the TCP/IP layer [8]. CZ.NIC, the Czech domain registry, implemented and slightly modified the algorithm for the purpose of DNS traffic monitoring [18]. They implemented a set of modules, called policies, in order to test various packet identifiers. Such a policy defines which packet features are used as the hashing keys for dividing packets into sketches. CZ.NIC designed and implemented[1] the first two policies, we added the third.

1. **IP Address Policy** [18] uses only the source IP address as the hash key. It is based on the original algorithm, which uses the whole TCP/IP connection identifier (source IP, destination IP, source port, destination port) as the hash key. However, destination address and destination port show little to no variability in TLD DNS traffic.
2. **Query Name Policy** [18] is based on application layer data. It extracts the first domain name from the DNS query and uses it as the hash key.
3. **ASN Policy** is a generalization of the *IP Address Policy*. For each source IP, it first looks up its autonomous system number (ASN). Each network on the Internet is uniquely defined by such an ASN. Therefore, this policy should be able to detect anomalies which are linked to a network, rather than an individual server.

3.3 QLAD-global

QLAD-flow will fail to detect attacks that use random spoofed IP addresses, since each packet will belong to a different flow. Especially DoS attacks often use this technique and will therefore remain undetected by *QLAD-flow*. *QLAD-global* tries to address this issue by looking at the global traffic flow.

The method is based on the observation that all common traffic anomalies cause changes in the distribution of one or more traffic features. In some cases, feature distributions become more concentrated on a small set of values; for example the distribution of source IP addresses during a reflection attack on a couple of servers in the same network. In other cases, feature distributions become more dispersed; for example when source addresses are spoofed in a DoS attack, or during a zone enumeration attack when lots of domains are queried only once. An analysis based on these traffic feature distributions can capture fine-grained patterns in traffic distributions that simple volume based metrics cannot identify [15].

---

[1] Project repository can be found at git://git.nic.cz/dns-anomaly/.

In this paper, we focus on eight traffic features: the number of requests for each TLD and SLD (second level domain name), the query types, the response codes, the number of requests by each client, ASN and country, and finally the response sizes. Obviously, these are not the only fields that could be used to detect anomalies, but they are general enough to encompass most other fields and we found that they suffice to detect the most common anomalies.

The distribution of these traffic features is high dimensional, and so is hard to work with directly. Therefore, we perform anomaly detection on the entropies of these feature distributions instead. The entropy is defined as:

$$H(X) = -\sum_{i=1}^{n} p(x_i) \log(p(x_i)).$$

Here, X is a feature that can take values $\{x_1, \ldots, x_n\}$ and $p(x_i)$ is the probability mass function of outcome x_i. Entropy is a measure of the uncertainty associated with a random variable [24]. Although, in this context, it can be interpreted as a measure for the degree of concentration or dispersal of a distribution [15].

Besides concentration and dispersal of the underlying distribution, the entropy also depends on the number of distinct values n. In practice, this means that entropy tends to increase when the traffic volume increases [15]. This is an advantage if normal traffic volumes are constant, since it allows the detection of volume anomalies with entropy measures. However, in our application, traffic volume is highly periodic, so we mitigate the effect of this phenomenon by normalizing entropy values. Therefore, we divide by $\log(n)$ (the maximum entropy), as proposed by Nychis et al. [22].

Figure 1 illustrates that entropy can be effective for anomaly detection with a simple example. The left plot shows the distribution between queries for valid (NOERROR) and non-existent domain names (NXDOMAIN) during 100 successive one-minute windows. The right plot shows the entropy values for these distributions. During two short periods (around minute 90 and around minute 125), the percentage of NXDOMAIN requests increases significantly, i.e., the distribution of response code values becomes more dispersed. These anomalies cannot be observed in the global traffic volume, but they stand out clearly on the right plot. The entropy increases, corresponding to a distributional dispersion around the NOERROR and NXDOMAIN response codes.

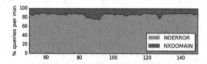
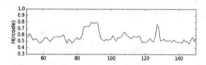

Fig. 1. Two anomalies in the ratio between DNS queries for valid and non-existent domain names over time (left), viewed in terms of entropy (right). Transforming the distribution to an entropy time series allows automated detection with standard anomaly detection techniques.

For each of the eight traffic features, we compute the entropy of the distribution in one-minute windows. This reduces the problem of anomaly detection in DNS traffic to a problem of anomaly detection in (correlated) time series. The research community has defined a wide range of algorithms to solve this type of problem [29]. One popular approach compares windows or individual points with a pre-trained profile of legitimate traffic [5]. Due to the lack of clean training data, we can not apply this approach to our problem. Alternatively, the subspace method [15] uses principal component analysis to detect unusual variations in the correlation between multiple time series. This method gave poor results, due to the high level of periodicity and unequal effects of this periodicity on each entropy time series. Therefore, we apply anomaly detection separately on each time series. There exists algorithms that should be able to learn the periodicity in these time series, but since the time series are influenced by both the hour of the day and the day of the week and since the data is contaminated with anomalies, these algorithms would require data of several weeks to learn a good model [31].

Despite its simplicity, we found a simple Exponential Moving Average (EMA) filter to work best. To quantify the range around the EMA that could be considered normal, we compute the Exponential Moving Standard Deviation (EMS):

$$\text{EMS}_t = \sqrt{w * \text{EMS}_{t-1}^2 + (1 - w) * (y_t - \text{EMA}_t)^2} \tag{1}$$

An anomaly is reported when $|y_t - \text{EMA}_t| > n * \text{EMS}_t$, where n controls the sensitivity of the alarm to the entropy measurement y_t.

4   QLAD Architecture

Our goal is to develop an anomaly detection system that is tuned to the unique nature of DNS traffic. Ideally, the system has the following characteristics:

Accurate. It should be able to detect suspicious and unexpected behaviour. This requires being sensitive to a wide range of possible attacks, including low volume anomalies.

Efficient. It should be able to handle high volumes of traffic and perform its analysis in near real time.

Unsupervised. It does not need any initial knowledge about the analysed traffic in the form of labelled data or a description of the possible attack vectors.

Provide insight. It should allow the operator to pinpoint the cause of an anomaly and determine the correct countermeasures.

Extendable. The architecture should enable future growth and improvements to the system.

The *QLAD* system was designed with these requirements in mind. Figure 2 displays its modular architecture, which can be divided into three layers: a data transformation, anomaly detection, and presentation layer. The transformation layer is a crucial component. Storing the raw pcap files is undesirable because they have a large storage cost and are inefficient to query. Therefore, we considered two alternative solutions for preprocessing and storing these pcap files.

The first solution, based on the DNS Statistics Collector (DSC) [9], transforms the pcap files into time aggregated traffic features such as the number of queries by query type, the most popular domains queried, and the length of DNS reply messages. The aggregated data is small enough that it can be directly stored in a traditional database system. This means that this component can be deployed easily and at a very low cost in terms of infrastructure. The aggregated data suffices to detect anomalies, but does not allow a detailed manual analysis.

The alternative, ENTRADA [32], is a big data platform designed to ingest and quickly analyse large amounts of network data. It was built by SIDN Labs, the research department of the Dutch domain registry, to enable fast data analysis on the network traffic of their authoritative name servers. ENTRADA is built entirely on open-source tools, and was open sourced itself in January 2016 [26]. The ENTRADA platform continuously processes pcap files in order to convert them to Apache Parquet files (a columnar storage format), which are ultimately stored in a Hadoop cluster (HDFS). Once archived, these files are available for analysis via Impala[2] [25].

Each approach has its own advantages and disadvantages. Effectively, the choice between both methods is a trade-off between deployment costs and more extensive data analysis possibilities.

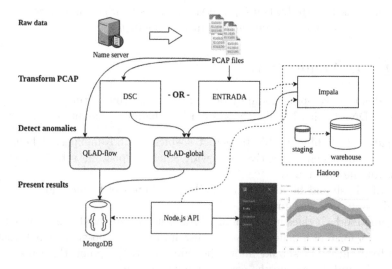

Fig. 2. A global overview of the *QLAD* system.

---

[2] Impala provides an SQL-like interface to query data stored in a HDFS. See https://impala.incubator.apache.org/overview.html for more information.

5   Empirical Evaluation

We evaluated *QLAD* on two days of real-world traffic to one of DNS Belgium's servers. The absence of a ground truth makes evaluating anomaly detectors notoriously difficult [27]. Since manually identifying all true anomalies in the traffic is infeasible, it is impossible to determine which anomalies are missed by the detectors. Hence, we cannot compute metrics like recall. The research community takes two approaches to address this problem [14]. A first approach is to manually inspect each detected anomaly in order to pinpoint its root cause. This approach fits best with a real-world application of the detector. A second approach injects the data with synthetically generated attacks. This approach has the advantage that the parameters of an attack (duration, traffic volume, etc.) can be carefully controlled, which enables sensitivity analysis of the detection algorithm.

With *QLAD* we focussed on the detection of a wide range of anomalies. Our primary interest is to find out which type of real-world anomalies *QLAD* is able to detect. Therefore, we use the first approach. An evaluation with the second approach is left as future work. Specifically, we address the following questions:

Q1: What are the space/time trade-offs between DSC and ENTRADA storage?
Q2: What is the runtime performance of the anomaly detectors?
Q3: Which types of real-world anomalies can *QLAD* detect?

5.1   Q1: Data Preprocessing and Storage

The traffic used for evaluation originates from one of DNS Belgium's name servers, collected between Sunday 12 February and Monday 13 February 2017. Traffic was captured for 44 h and amounts to 58,345,819 queries or 42 GB of pcap packet dumps. We processed these logs with both ENTRADA and DSC. ENTRADA reduces the original logs to 3.8 GB of Parquet files, which is about 9% of the original volume. Note, however, that—given the default replication factor of three in Hadoop—you still need 11.4 GB of actual storage space. With DSC, the volume is reduced to 39 MB, which is about 0.09% of the original size.

5.2   Q2: Anomaly Detection Performance

Both *QLAD-flow* and *QLAD-global* are computationally very efficient and capable of analysing real-time traffic. *QLAD-flow* analyses the complete dataset in 3 m 36 s (source IP policy), 5 m 13 s (query name policy) and 21 m 45 s (ASN policy).[3] We used a window size of 10 min, 8 aggregation levels, 25 hash functions, a hash table of size 32 and 1.2 as a detection threshold. The same parameters are used in further experiments. For a more extensive performance analysis with varying parameters, we refer to Mikle et al. [18].

Using the DSC setup, *QLAD-global* can analyse the full dataset in 1 m 18 s. Its performance is mainly determined by the time needed to fetch the data from

---

[3] 2.3 GHz dual-core Intel Core i5 processor with 8 GB RAM.

the MongoDB database. We do not count the time needed to extract the traffic feature distributions from the query logs.

5.3   Q3: Classification of Detected Anomalies

Table 1 shows the total number of anomalies detected by each anomaly detection algorithm and our expert's classification. We tried to group anomalies together as much as possible. For example, when the same attack is launched from multiple IP addresses, we count it as only one anomaly. Furthermore, there was a large overlap between the anomalies detected by each detector. We write the number of anomalies that are not detected by any other detection method in parentheses. *QLAD-flow* with the ASN policy detected a subset of the anomalies detected with the source IP policy, but no new anomalies.

Table 1. A manual classification of the detected anomalies.

Anomaly type	QLAD-flow (source IP)	QLAD-flow (query name)	QLAD-global	Total
Benign				
Caching resolver	12 (10)		2	12
Email marketing	8 (7)		2 (1)	9
Other	1 (1)	2 (2)		3
Malicious				
Spam sender	3 (3)			3
Domain enumeration	5 (3)		2	5
Reflection attack	1		1	2
Phishing	1 (1)			1
DoS attack	3 (1)	2 (1)	1	4
Unknown	1		1	1
Total	35 (26)	4 (3)	9 (1)	39

In total, we found 39 unique anomalies, but only a minority of them is interesting for a TLD operator. *QLAD-flow* pinpoints flows which are statistically different from the other flows in some time window. Not all of these flows are malicious; there are some benign applications with unique traffic patterns too. Although, they are anomalous in some sense, they are not interesting for a TLD operator. Examples of such applications we found include caching resolvers, email marketing services, a cloud provider, security appliances and a cryptocurrency mining pool.

Other anomalies were clearly malicious. We found something that looked like a reflection attack on Twitter, several domain enumeration attacks, three servers sending spam and four DoS attacks (although not large enough to have a real impact). Furthermore, one server generated bizarre queries for domains such as www-axa-fr.be and www-keytradebank.be. Since these refer to Belgian and French banks, these queries where probably generated by a script looking for unused domain names to be used in phishing.

To conclude, we verified whether any of the IP addresses corresponding to the anomalies we discovered was reported previously for engaging in malicious activity.[4] Two were reported for engaging in DDoS attacks and two others for spreading spam. For all four, this corresponds to the anomalies we observed.

6   Conclusions

We presented the design, implementation and evaluation of the $QLAD$ framework—a proof of concept anomaly detection system for query logs of a TLD resolver. We showed how existing solutions can be integrated with custom development, to create a system that is adapted to the particular nature of DNS traffic at a TLD resolver and that can detect a wide range of anomalies in near real-time.

Acknowledgements. The authors acknowledge the partial support of KU Leuven Research Fund C14/17/070 and C22/15/015 (PR and JD), FWO-Vlaanderen SBO-150033 (JD and WM) and Interreg V A project NANO4Sports (PR and JD).

References

1. Alonso, R., Monroy, R., Trejo, L.: Mining IP to domain name interactions to detect DNS flood attacks on recursive DNS servers. Sensors **16**(8), 1311 (2016)
2. Ariyapperuma, S., Mitchell, C.J.: Security vulnerabilities in DNS and DNSSEC. In: Proceedings of the 2nd International Conference on Availability, Reliability and Security, pp. 335–342. IEEE Computer Society, Washington, DC (2007)
3. Baker, M.: Packetpig - open source big data security analysis. http://blog.packetloop.com/2012/03/packetpig-open-source-big-data-security.html
4. Bär, A., Finamore, A., Casas, P., Golab, L., Mellia, M.: Large-scale network traffic monitoring with DBStream, a system for rolling big data analysis. In: 2014 IEEE International Conference on Big Data, pp. 165–170, October 2014
5. Bereziński, P., Jasiul, B., Szpyrka, M.: An entropy-based network anomaly detection method. Entropy **17**(4), 2367–2408 (2015)
6. Bhuyan, M.H., Bhattacharyya, D.K., Kalita, J.K.: Network anomaly detection: methods, systems and tools. IEEE Commun. Surv. Tutor. **16**(1), 303–336 (2014)
7. Cheng, T.H., Lin, Y.D., Lai, Y.C., Lin, P.C.: Evasion techniques: sneaking through your intrusion detection/prevention systems. IEEE Commun. Surv. Tutor. **14**(4), 1011–1020 (2012)
8. Dewaele, G., Fukuda, K., Borgnat, P., Abry, P., Cho, K.: Extracting hidden anomalies using sketch and non Gaussian multiresolution statistical detection procedures. In: Proceedings of the 2007 Workshop on Large Scale Attack Defense, pp. 145–152. ACM, New York (2007)

---

[4] We used the AbuseIP database (https://www.abuseipdb.com).

9. DNS-OARC: The DNS Statistics Collector, February 2016. https://www.dns-oarc.net/tools/dsc

10. Gascon, H., Orfila, A., Blasco, J.: Analysis of update delays in signature-based network intrusion detection systems. Comput. Secur. **30**(8), 613–624 (2011)

11. Hesselman, C., Moura, G.C.M., Schmidt, R.d.O., Toet, C.: Increasing DNS security and stability through a control plane for top-level domain operators. IEEE Commun. Mag. **55**(1), 197–203 (2017)

12. Karasaridis, A., Meier-Hellstern, K., Hoeflin, D.: NIS04-2: detection of DNS anomalies using flow data analysis. In: IEEE Globecom 2006, pp. 1–6 (2006)

13. Kreibich, C., Warfield, A., Crowcroft, J., Hand, S., Pratt, I.: Using packet symmetry to curtail malicious traffic. In: Proceedings of the 4th Workshop on Hot Topics in Networks (2005)

14. Lakhina, A., Crovella, M., Diot, C.: Diagnosing network-wide traffic anomalies. SIGCOMM Comput. Commun. Rev. **34**(4), 219–230 (2004)

15. Lakhina, A., Crovella, M., Diot, C.: Mining anomalies using traffic feature distributions. SIGCOMM Comput. Commun. Rev. **35**(4), 217–228 (2005)

16. Lee, Y., Lee, Y.: Toward scalable internet traffic measurement and analysis with hadoop. SIGCOMM Comput. Commun. Rev. **43**(1), 5–13 (2012)

17. Li, Z., Das, A., Zhou, J.: USAID: unifying signature-based and anomaly-based intrusion detection. In: Ho, T.B., Cheung, D., Liu, H. (eds.) PAKDD 2005. LNCS (LNAI), vol. 3518, pp. 702–712. Springer, Heidelberg (2005). https://doi.org/10.1007/11430919_81

18. Mikle, O., Slany, K., Vesely, J., Janousek, T., Survy, O.: Detecting hidden anomalies in DNS communication. Technical report, CZ.NIC (2011)

19. Mockapetris, P.: Domain names - concepts and facilities. STD 13, RFC Editor, November 1987. http://www.rfc-editor.org/rfc/rfc1034.txt

20. Newman, L.H.: What we know about Friday's massive internet outage, October 2016. http://www.wired.com/2016/10/internet-outage-ddos-dns-dyn/

21. NOMINET: NOMINET adds machine learning to Turing network analytics and monitoring tool, February 2017. https://www.nominet.uk/nominet-adds-machine-learning-to-turing-network-analytics-and-monitoring-tool/

22. Nychis, G., Sekar, V., Andersen, D.G., Kim, H., Zhang, H.: An empirical evaluation of entropy-based traffic anomaly detection. In: Proceedings of the 8th ACM SIGCOMM Conference on Internet Measurement, pp. 151–156 (2008)

23. Satam, P., Alipour, H., Al-Nashif, Y., Hariri, S.: Anomaly behavior analysis of DNS protocol. JISIS **5**(4), 85–97 (2015)

24. Shannon, C.E.: A mathematical theory of communication. SIGMOBILE Mob. Comput. Commun. Rev. **5**(1), 3–55 (2001)

25. SIDN Labs: ENTRADA Documentation. http://entrada.sidnlabs.nl/docs/introduction/overview/

26. SIDN Labs: SIDN Labs Open-Sources ENTRADA. https://www.sidnlabs.nl/a/weblog/sidn-labs-open-sources-entrada

27. Silveira, F., Diot, C., Taft, N., Govindan, R.: ASTUTE: detecting a different class of traffic anomalies. In: Proceedings of the ACM SIGCOMM Conference, pp. 267–278 (2010)

28. Son, S., Shmatikov, V.: The Hitchhiker's guide to DNS cache poisoning. In: Jajodia, S., Zhou, J. (eds.) SecureComm 2010. LNICSSITE, vol. 50, pp. 466–483. Springer, Heidelberg (2010). https://doi.org/10.1007/978-3-642-16161-2_27

29. Teng, M.: Anomaly detection on time series. In: 2010 IEEE International Conference on Progress in Informatics and Computing, vol. 1, pp. 603–608 (2010)

30. Trostle, J., Van Besien, B., Pujari, A.: Protecting against DNS cache poisoning attacks. In: 6th IEEE Workshop on Secure Network Protocols, pp. 25–30 (2010)
31. Vallis, O., Hochenbaum, J., Kejariwal, A.: A novel technique for long-term anomaly detection in the cloud. In: Proceedings of the 6th USENIX Conference on Hot Topics in Cloud Computing, pp. 15. USENIX Association, Berkeley (2014)
32. Wullink, M., Moura, G.C.M., Muller, M., Hesselman, C.: ENTRADA: a high-performance network traffic data streaming warehouse. In: 2016 IEEE/IFIP Network Operations and Management Symposium, pp. 913–918. IEEE, April 2016
33. Yuchi, X., Wang, X., Lee, X., Yan, B.: A new statistical approach to DNS traffic anomaly detection. In: Cao, L., Zhong, J., Feng, Y. (eds.) ADMA 2010. LNCS (LNAI), vol. 6441, pp. 302–313. Springer, Heidelberg (2010). https://doi.org/10.1007/978-3-642-17313-4_30

Multimodal Tweet Sentiment Classification Algorithm Based on Attention Mechanism

Peiyu Zou[1(✉)] and Shuangtao Yang[2(✉)]

[1] Northeast Agricultural University, Harbin 150036, China
zoupeiyu1213@gmail.com
[2] Lenovo AI Lab, Beijing 100000, China
lufiedby@gmail.com

Abstract. With the rapid development of Internet, multimodal sentiment classification has become an important task in natural language processing research. In this paper, we focus on the sentiment classification of tweets that contains both text and image, a multimodal sentiment classification method for tweets is proposed. In this method Bidirectional-LSTM model is used to extract text modality features and VGG-16 model is used to extract image modality features. Where all features are extracted, a new multimodal feature fusion algorithm based on attention mechanism is used to finish the fusion of text and image features. This fusion method proposed in this paper can give different weights to modalities according to their importance. We evaluated the proposed method on the Chinese Weibo dataset and SentiBank Twitter dataset. The experimental results show method proposed in this paper is better than models that only use single modality feature, and attention based fusion method is more efficient than directly summing or concatenating features from different modalities.

Keywords: Multimodal · Sentiment classification · Attention mechanism

1 Introduction

With the rapid development of Internet, social network become more and more popular in daily live. People begin to tell others what they are doing, what they are feeling, what they are thinking, or what is happening around them more and more through social network. As a result, how to excavate people's sentiment expression in social networks exactly has attracted more and more attention. Sentiment classification has become an important task in the natural language processing research.

Some related works have been conducted on social network (Twitter or Sina Weibo and so on) sentiment classification, such as [1–4], while these work mainly focus on the use of single modality feature for sentiment classification, most mainly use text features, and rarely use image, audio or visual feature. With the rapid development of social network, user's tweets are often multimodal, when users publish tweets, may upload an image, audio or video at the same time. If we cannot effectively use these multimodal features, we will not be able to accurately evaluate user's sentiment. So,

A. Monreale et al. (Eds.): ECML PKDD 2018 Workshops, CCIS 967, pp. 68–79, 2019.
https://doi.org/10.1007/978-3-030-14880-5_6

recently multimodal sentiment classification has attracted more and more attention, such as [5–8].

In this paper, we will focus on the sentiment classification of tweets (also include micro-blogs from Sina Weibo) that contain both text and image. A sentiment classification method for multimodal tweets classification is proposed, which use BI-LSTM model and VGG-16 model to extract text features and image features separately and then use attention mechanism to complete the fusion of these two features. The proposed method has been tested on Chinese Weibo dataset and SentiBank Twitter dataset, all showing promising results.

This paper is organized as follows: Sect. 2 introduces related works of multimodal sentiment classification. Section 3 describes the method proposed in this paper. Section 4 carries out some comparison experiment and result analysis. And conclusions are shown in Sect. 5.

2 Related Work

Multimodal Sentiment Classification can be divided into two categories: early fusion and late fusion according to modality information fusion time as described in [9, 10]. In early fusion (also named feature fusion), features of different modalities are first extracted by different feature extraction model, then these extracted features are fused, which can be finished by concatenating, summing or by some fusion models. Finally, the fusion features are fed into classifier. Similar to the early fusion strategy, in the late fusion (also named decision fusion), different models are used to extract features from different modalities, but late fusion does not fuse features directly, it will first feed features to a classifier in each modality, and finally a model will be used to combine all classification results from different modalities to get the final classification results.

In multimodal sentiment classification, some related works have been conducted. Paper [5] explored the joint use of multiple modalities for the purpose of classifying the polarity of opinions in online videos and experiment result showed that the integration of visual, audio, and textual features can improve significantly over the individual use of one modality at a time. In [6] and [11], Support Vector Machine (SVM) is used for multimodal sentiment classification, they both first extract features from different modalities, then combine features into a single vector and feed the vector into SVM classifier. In [7], a unified model (CBOW-DA-LR) was proposed, which works in an unsupervised and semi-supervised way to learn text and image representation. For video sentiment classification task, in order to capture inter-dependencies and relations among the utterances in a video, [12] developed a LSTM-based network to extract contextual features from the utterances and got better result comparing to traditional method. [8] proposed a novel method for multimodal emotion recognition and sentiment analysis, which uses deep convolutional neural networks to extract features from visual and textual modalities and then feeds such features to multiple kernel learning (MKL) classifier which is a feature selection method and it is able to combine data from different modalities effectively. Paper [13] proposed a multimodal affective data analysis framework which can extract user opinion and emotions from video content, and multiple kernel learning is also used to combine visual, audio and textual modalities. Paper [14] introduced a new end-to-end early fusion method for multimodal sentiment analysis termed Tensor Fusion Network which use a fusion layer to disentangles uni-modal, bimodal and tri-modal dynamics by modeling each of them explicitly. Paper [15] proposed a multimodal sentiment analysis model named Select-Additive Learning which attempts to prevent identity-dependent information from being learned in a deep neural network. Paper [16] proposed a model named GME-LSTM which is able to better model the multimodal structure of speech through time and perform better sentiment comprehension. GME-LSTM is composed of 2 modules: Gated Multimodal Embedding which alleviates the difficulties of fusion when there are noisy modalities; LSTM performs word level fusion at a finer fusion resolution between input modalities and attends to the most important time steps.

3   The Proposed Model

In this paper, we will focus on early fusion strategy. According to the process of early fusion strategy, we can find that it attempts to fuse the features from different modalities to obtain a global feature which contains all valuable features from every modality. In the fusion, the simplest approach is to concatenate or sum features from different modalities directly to get the global representation. Considering concatenate or sum fusion method does not reflect the importance of feature from different modalities, in this paper, we propose a multimodal feature fusion method based on attention mechanism, which will give different weights to different modalities according to their importance, and it helps to complement and disambiguate different features from different modalities. Our model is shown as Fig. 1.

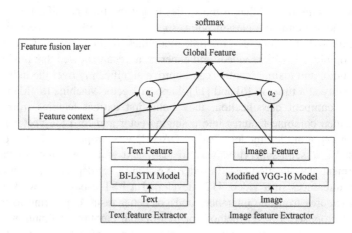

Fig. 1. Attention based multimodal sentiment classification model

3.1   Text Feature Extractor

In order to get better text modal features, it is necessary to select the appropriate text feature (lexical) representation method first. Text feature representation has been extensively studied in text categorization tasks. Mainstream text representation methods include word bag representation and word vector representation. In word-bag model word is represented in one-hot vector which cannot effectively express the relation of different words, for example "like" and "love". In addition, the dimension of the one-hot vector needs to be consistent with the size of the vocabulary. The dimension one-hot vector is often too high and extremely sparse which is not unfriendly to computation. So, text features will be represented by word2vec in this paper.

In word2vec, all words are embedded into an N-dimension semantic space [17, 18]. In this semantic space, the semantic distance of related words will be closer, and the semantic distance is far away from unrelated words. After the words are represented into vectors, the sentence will be transformed into M * N vector, while N is the dimension of word2vec and M is the word number of the sentence. Then we can use CNN or LSTM model to encode this two-dimensional vectors to obtain effective text modal features [19].

Using CNN or LSTM to encode the sentence, it can effectively capture the sequence information between words in the sentence [20–23]. In this paper we will use Bi-directional LSTM to extract text feature. BI-LSTM can be considered as a composed of two different directions LSTM, mainly to compensate for the deficiency of the single direction LSTM in capturing context ability. The process of text feature extraction based on Bi-LSTM model is shown in the following Fig. 2.

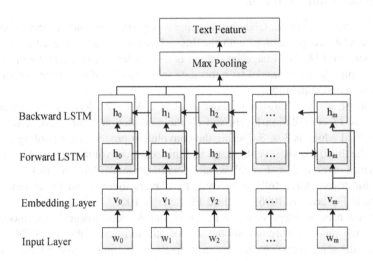

Fig. 2. Bidirectional-LSTM text feature extractor

Given a sentence with words w_t $t \in [0, m]$, we first transfer every word w_t to a N-dimension vector v_t through a pre-trained word2vec. Then we will use LSTM to encode sentence from both directions, as shown in Fig. 2.

In forward LSTM, sentence is encoded in the direction from word w_0 to word w_m. At t time-step, the hidden layer state of the LSTM is expressed as:

$$f\_h_t = forward\_lstm(v_t), \ t \in [0, m] \tag{1}$$

In backward LSTM, sentence is encoded in the direction from word w_m to word w_0. At t time-step, the hidden layer state of the LSTM is expressed as:

$$b\_h_t = backward\_lstm(v_t), \ t \in [0, m] \tag{2}$$

We will get h_t for a given word w_t by concatenating the forward hidden state f_h_t and forward hidden state b_h_t.

$$h_t = concatenate(f\_h_t, b\_h_t) \tag{3}$$

Then we need run max pooling over sequence $[h_0, h_1, \ldots, h_m]$, and get the final feature. In order to make it easy to explain, we set the dimension of LSTM hidden layer in both forward LSTM and backward LSTM to d, thus for each sentence we're going to get a 256-dimensional vector as its text modality feature. Considering CNN models are also widely used in NLP tasks, we will also evaluate the performance of famous Text-CNN model [20] in our text feature extraction task.

3.2   Image Feature Extractor

In image feature extraction, CNN has become a very standard and universal model. In this paper VGG-16 pre-trained with the ImageNet dataset will be used as an image feature extractor [24, 25]. ImageNet dataset is a super large image annotation data set, and pre-training VGG-16 model with this data set is very helpful to obtain better image feature extraction capability.

The input of the VGG-16 network is $3 \times 224 \times 224$ RGB color image. The network has 13 convolution layers and all the convolution layers have a very small receptive field which is 3×3. During the convolution process, max pooling is carried out after the third, fourth, seventh, tenth and thirteenth convolution layer. Max pooling is performed over a 2×2 pixel window with stride 2. After a stack of thirteen convolution layers, three fully-connected (FC) layers follows and the channel of the first, second FC layer is 4096, the third FC layer is 1000.

First, we need to preprocess the image before we feed image to the model. The preprocessing phase mainly adjusts the size of all the image data to $3 \times 244 \times 244$, and we do the same preprocessing for each pixel: subtracting the mean RGB value for each pixel in the training set. The images are enhanced by random transformation in training set.

Second, we pre-train the VGG-16 model on Image Net dataset. When we finish the pre-training, model parameters except the last full connected layer in the model are frozen. We will modify the output dimension of last layer to $2 * d$, and d is the dimension of LSTM hidden layer which has been introduced in Sect. 3.1. In the proposed model, image features and text features need to be mapped into same dimension space which means the dimension of the text feature should be consistent with the dimension of text feature. Therefore, we need to transform the dimension of the image features through the last FC layer which dimension should be $2 * d$.

3.3   Attention Base Fusion of Text and Image Feature

In early fusion, after we get text features and image features, we have to fusion them to get global features which will contain all important features from different modalities. We believe that not all features from different modalities contribute equally to the global feature. In order to get better feature fusion results, we propose an attention mechanism based fusion method for text and image features. In this method we will pay more attention to the modalities that are more important to classification which means features extracted from the modalities will contribute more in global feature.

For illustrative purposes, we record the textual features extracted by Text Feature Extractor (introduced in Sect. 3.1) f_{text}, and mark the image features extracted by Image re Extractor (introduced in Sect. 3.2) as f_{image}, and they have the same dimensions. First, we feed the f_{text} and f_{image} to a one-layer MLP respectively, we will get:

$$u_{text} = MLP(f_{text}) = \tanh(f_{text}W_u + b_u) \tag{4}$$

$$u_{image} = MLP(f_{image}) = \tanh(f_{image}W_u + b_u) \tag{5}$$

where W_u represents the weight matrix in MLP, and b_u represents the bias in MLP.

Then, we will measure the importance of text features and image features by calculating the similarity between the u_{text} and the u_{image} and $u_{context}$, parameter $u_{context}$ is randomly initialized and jointly learned during the training process. We also will use softmax function to normalize the feature weights a_{text} and a_{image}. The attention score a_{text} for text feature is computed as formula, where u_{text}^T is the transpose of u_{text}, and u_{image}^T is the transpose of u_{image}.

$$a_{text} = \frac{\exp\left(u_{text}^T u_{context}\right)}{\exp\left(u_{text}^T u_{context}\right) + \exp\left(u_{image}^T u_{context}\right)} \tag{6}$$

The attention score a_{image} for image feature is computed as:

$$a_{image} = \frac{\exp\left(u_{image}^T u_{context}\right)}{\exp\left(u_{text}^T u_{context}\right) + \exp\left(u_{image}^T u_{context}\right)} \tag{7}$$

Then we can get global_feature as shown in formula 8.

$$global\_feature = a_{text}f_{text} + a_{image}f_{image} \tag{8}$$

4   Experiment

4.1   Datasets

(1) Chinese Weibo Dataset

We first collected lots of tweet (or named Weibo) from Sina Weibo which is a famous social network in China. Then 12000 tweets are annotated manually. Every tweet contains text and image. All the tweets are labeled as three categories: Positive, Negative and Neutral. The categories of this dataset are balanced, each category, and each category contains 4000 annotated tweets. In order to verify the effectiveness of the proposed model, we adopt a 5-fold cross validation method, which means in each validation 80% annotated tweets are used as training set and 20% are used as testing set. The training set and test set are preprocessed, for text, we mainly normalized the special fields, such as emotion tags and URL, and used Jieba (a Chinese word segmentation tool) to segment tweets content into words.

(2) SentiBank Twitter Dataset

SentiBank Twitter Dataset consists of 470 positive and 133 negative tweets with images, related to 21 topics, annotated using Mechanical Turk [26]. For this dataset, we also use 5-fold cross-validation to evaluate our model.

4.2   Result and Analysis

It needs to be pointed out that we used the pre trained word vectors, respectively. For Chinese Weibo Dataset we pre-trained a 300 dimension word2vec on a very large Chinese text corpus. For SentiBank Twitter Dataset, we also pre-trained a 300 dimension word2vec on other twitter corpus and wiki corpus. We will use micro F1-score to evaluate our model, and we calculate it as described in paper [27]:

$$micro\_precision = \frac{\sum_{j=1}^{Q} tp_j}{\sum_{j=1}^{Q} tp_j + \sum_{j=1}^{Q} fp_j} \tag{9}$$

where Q is the total number of testing case, tp_j and fp_j are the number of true positives and false positives for the label j_{label} considered as a binary label.

$$micro\_recall = \frac{\sum_{j=1}^{Q} tp_j}{\sum_{j=1}^{Q} tp_j + \sum_{j=1}^{Q} fn_j} \tag{10}$$

where fn_j is the number of false negatives for the label j_{label} considered as a binary label.

$$micro\_f1 = \frac{2 \times micro\_precision \times micro\_recall}{micro\_precision + micro\_recall} \qquad (11)$$

First, we compared the model proposed in this paper with models those can only use single modality feature (text feature or image feature only). Experiment results are shown in Table 1.

Table 1. Result on Chinese Weibo dataset

Feature	Model	Micro F1-score
Text (only)	CBOW+SVM	71.9%
	Bi-LSTM	73.9%
	Text-CNN	72.7%
Image (only)	VGG-16-modified	70.1%
Text + image	Bi-LSTM+VGG-16	84.2%
	Text-CNN+VGG-16	82.9%

When only use text modality features, we carried out three classification experiments of CBOW+SVM model, BI-LSTM model and Text-CNN model respectively. We finally find that Bidirectional-LSTM model gets the best classification result, comparing to the Text-CNN model, there is a 1.2% increase of micro f1-score.

For image modality experiment, we evaluate VGG-16 modified model which has been described in Sect. 3.2 and get score of 71.5%. By comparing the features of the text separately and using the image features alone, we can find that the text feature contains more explicit sentiment information and is easier to get better classification results. As shown in Table 1, when using text feature and image feature both, we can find the model proposed in the paper gets the best classification result, and comparing to models that using single modality feature there is a particularly obvious improvement, and the micro f1-score is up to 84.2%. As shown in Table 1, we also use Text-CNN model as text feature extractor to replace Bidirectional-LSTM model in our model, and get average accuracy of 82.9%. Comparing to Bi-direction-LSTM version, the micro f1-score drops by 1.3% (Tables 2 and 3).

Table 2. Result on SentiBank dataset

Feature	Model	Micro-F1 score
Text (only)	CBOW+SVM	72.4%
	Bidirectional-LSTM	73.8%
	Text-CNN	73.3%
Image (only)	VGG-16-modified	71.2%
Text + Image	BI-LSTM+VGG-16	82.8%
	Text-CNN+VGG-16	80.7%

Table 3. Result on Chinese Weibo dataset

Feature	Fusion method	Micro-F1 score
Text + image	Sum	79.6%
	Concatenate	82.3%
	This paper	84.2%

For SentiBank Twitter Dataset we conducted the same comparative experiment, and the specific results are as follows:

When only using text modality feature, Bidirectional-LSTM also achieved the highest micro f1-score among CBOW+SVM model, Bidirectional-LSTM model and Text-CNN model. If text modality feature and image feature are all used, our model also achieved the best result.

In multimodal comparison experiments, we also compared the influence of different features (text feature and image feature) fusion methods on sentiment classification. The first fusion method is sum, which means the global feature is the sum of text feature f_{text} and image feature f_{image}, using this fusion method micro f1-score is 79.6%. The second fusion method is concatenate, which means the global feature is the concatenation of text feature f_{text} and image feature f_{image}, and the micro f1-score is 82.3%. The third fusion method is the attention based fusion method proposed in the paper, it achieve the highest micro f1-score which is up to 84.2%.

SentiBank experiments can be found in Table 4. The results also show that the feature fusion method based on attention mechanism is more effective than simple sum or concatenate features from different modalities.

Table 4. Result on SentiBank dataset

Feature	Fusion method	Micro-F1 score
Text + Image	Sum	79.6%
	Concatenate	81.2%
	This paper	82.1%

In order to better explain our method, we find some illustrative examples from Weibo Dataset where are shown in Table 5. For the first case, if we predict only based on text modality features, we get negative sentiment which is wrong. When we use text features and image features both, we get the right sentiment label, and the attention score of text modality is 0.29, which is much lower than that of image modality. In the third case, there is a man who sprained his ankle in the picture, so if we only use image modality features, we get negative sentiment, however when we focus on the text content we can make sure that the sentiment of this case is positive. As we can see, our attention based fusion model also paid more attention to text modality features and the attention score of text modality is 0.84, which is much higher than image modality.

Table 5. Illustrative examples from Weibo testset

Image and Text	Feature	Classification result
Even life is always disappointing and frustrating, we can't stop. (Translated from Chinese)	Text	Negative Confidence = 0.82
	Image	Positive Confidence = 0.35
	Text+Image	Positive Confidence = 0.78 Text modality attention score: 0.29 Image attention score: 0.71
		Ground truth: Positive
Maybe god knows what kind of pain I had last night. (Translated from Chinese)	Text	Negative Confidence = 0.91
	Image	Positive Confidence = 0.61
	Text+Image	Positive Confidence = 0.76 Text modality attention score: 0.35 Image attention score: 0.65
		Ground truth: Positive
My hero, I hope you can recover soon, love, love and love. (Translated from Chinese)	Text	Positive Confidence = 0.96
	Image	Negative Confidence = 0.45
	Text+Image	Positive Confidence = 0.66 Text modality attention score: 0.84 Image attention score: 0.16
		Ground truth: Positive

5 Conclusion

In this paper, we focus on the sentiment classification of tweet that contains both text and image. A multimodal sentiment classification method for tweet is proposed. In this method Bidirectional-LSTM model is used to extract text modality feature and VGG-16 model is used to extract image modality feature. After all features are extracted, a new multimodal feature fusion algorithm based on attention mechanism is used to finish the fusion of text and image features. This fusion method proposed in this paper can give different weights to modalities according to its importance. We evaluated the proposed method on the Chinese Weibo dataset and SentiBank Twitter dataset. Our, experimental results show method proposed in this paper is better than models that use single modality feature, and attention based fusion method is more efficient than directly summing or concatenating features from different modalities.

References

1. Saif, H., He, Y., Alani, H.: Semantic sentiment analysis of Twitter. In: Cudré-Mauroux, P., et al. (eds.) ISWC 2012. LNCS, vol. 7649, pp. 508–524. Springer, Heidelberg (2012). https://doi.org/10.1007/978-3-642-35176-1_32
2. Gautam, G., Yadav, D.: Sentiment analysis of Twitter data using machine learning approaches and semantic analysis. In: International Conference on Contemporary Computing, pp. 437–442. IEEE (2014)
3. Zhou, H., et al.: Rule-based Weibo messages sentiment polarity classification towards given topics. In: Eighth SIGHAN Workshop on Chinese Language Processing, pp. 149–157 (2015)
4. Jiang, L., Yu, M., et al.: Target-dependent Twitter sentiment classification. In: Meeting of the Association for Computational Linguistics: Human Language Technologies, pp. 151–160. Association for Computational Linguistics (2011)
5. Morency, L.P., Mihalcea, R., Doshi, P.: Towards multimodal sentiment analysis: harvesting opinions from the web. In: International Conference on Multimodal Interfaces, pp. 169–176. ACM (2011)
6. Zadeh, A., Zellers, R., Pincus, E., Morency, L.P.: Multimodal sentiment intensity analysis in videos: facial gestures and verbal messages. IEEE Intell. Syst. **31**(6), 82–88 (2016)
7. Baecchi, C., Uricchio, T., Bertini, M., Bimbo, A.D.: A multimodal feature learning approach for sentiment analysis of social network multimedia. Multimedia Tools Appl. **75**(5), 2507–2525 (2016)
8. Poria, S., Chaturvedi, I., Cambria, E., et al.: Convolutional MKL based multimodal emotion recognition and sentiment analysis. In: IEEE, International Conference on Data Mining, pp. 439–448. IEEE (2017)
9. Atrey, P.K., Hossain, M.A., Saddik, A.E., Kankanhalli, M.S.: Multimodal fusion for multimedia analysis: a survey. Multimedia Syst. **16**(6), 345–379 (2010)
10. Gallo, I., Calefati, A., Nawaz, S.: Multimodal classification fusion in real-world scenarios. In: IAPR International Conference on Document Analysis and Recognition. IEEE (2018)
11. Pérez-Rosas, V., Mihalcea, R., Morency, L.P.: Utterance-level multimodal sentiment analysis. Association for Computational Linguistics (ACL) (2013)
12. Poria, S., Cambria, E., Hazarika, D., Majumder, N., Zadeh, A., Morency, L.P.: Context-dependent sentiment analysis in user-generated videos. In: Meeting of the Association for Computational Linguistics, pp. 873–883 (2017)
13. Poria, S., Peng, H., Hussain, A., Howard, N., Cambria, E.: Ensemble application of convolutional neural networks and multiple kernel learning for multimodal sentiment analysis. Neurocomputing **26**, 217–230 (2017)
14. Zadeh, A., Chen, M., Poria, S., Cambria, E., Morency, L.P.: Tensor fusion network for multimodal sentiment analysis (2017)
15. Wang, H., Meghawat, A., Morency, L.P., Xing, E.P.: Select-additive learning: improving cross-individual generalization in multimodal sentiment analysis, pp. 949–954 (2016)

16. Chen, M., Wang, S., Liang, P.P., Baltrušaitis, T., Zadeh, A., Morency, L.P.: Multimodal sentiment analysis with word-level fusion and reinforcement learning. In: ACM International Conference on Multimodal Interaction, pp. 163–171. ACM (2017)

17. Mikolov, T., et al.: Efficient estimation of word representations in vector space. In: International Conference on Learning Representations, pp. 1–12 (2013)

18. Mikolov, T., Sutskever, I., Chen, K., et al.: Distributed representations of words and phrases and their compositionality. In: International Conference on Neural Information Processing Systems, pp. 3111–3119. Curran Associates Inc. (2013)

19. Bengio, Y.: Learning deep architectures for AI. Found. Trends® Mach. Learn. **2**(1), 1–127 (2009)

20. Kim, Y.: Convolutional neural networks for sentence classification. Eprint Arxiv (2014)

21. Zhang, Y., Wallace, B.: A sensitivity analysis of (and practitioners' guide to) convolutional neural networks for sentence classification. Computer Science (2015)

22. Zhou, P., Qi, Z., Zheng, S., Xu, J., Bao, H., Xu, B.: Text classification improved by integrating bidirectional LSTM with two-dimensional max pooling (2016)

23. Chen, T., Xu, R., He, Y., et al.: Improving sentiment analysis via sentence type classification using BiLSTM-CRF and CNN. Expert Syst. Appl. **72**, 221–230 (2016)

24. Simonyan, K., Zisserman, A.: Very deep convolutional networks for large-scale image recognition. Computer Science (2014)

25. Deng, J., Dong, W., Socher, R., Li, L.J., Li, K., Li, F.F.: ImageNet: a large-scale hierarchical image database. In: IEEE Conference on Computer Vision and Pattern Recognition, CVPR 2009, pp. 248–255. IEEE (2009)

26. Bifet, A., Frank, E.: Sentiment knowledge discovery in twitter streaming data. In: Proceedings of the Discovery Science - International Conference, DS 2010, Canberra, Australia, 6–8 October 2010, pp. 1–15. DBLP (2010)

27. Madjarov, G., Kocev, D., Gjorgjevikj, D., Deroski, S.: An extensive experimental comparison of methods for multi-label learning. Pattern Recogn. **45**(9), 3084–3104 (2012)

Active Learning by Clustering for Drifted Data Stream Classification

Jakub Zgraja[1], João Gama[2], and Michał Woźniak[1(✉)]

[1] Department of Systems and Computer Networks,
Wrocław University of Science and Technology, Wrocław, Poland
{jakub.zgraja,michal.wozniak}@pwr.edu.pl
[2] Laboratory of Artificial Intelligence and Decision Support,
Faculty of Economics, University of Porto, Porto, Portugal
jgama@fep.up.pt

Abstract. Usually, during data stream classifier learning, we assume that labels of all incoming examples are available without any delay and they are used to update employing predictive model. Unfortunately, this assumption about access to all class labels is naive and it requires relatively high budget for labeling. It causes that methods which can train data stream classifiers on the basis of partially labeled data are highly desirable. Among them, active learning [1] seems to be a promising direction, which focuses on selecting only the most valuable learning examples to be labeled and used to produce an accurate predictive model. However, designing such a system we have to ensure that a chosen active learning strategy is able to handle changes in data distribution and quickly adapt to changing data distribution. In this work, we focus on novel active learning strategies that are designed for effective tackling of such changes. We propose a novel active data stream classifier learning method based on *query by clustering* approach. Experimental evaluation of the proposed methods prove the usefulness of the proposed approach for reducing labeling cost for classifier of drifting data streams.

Keywords: Active learning · Data streams · Classification

1 Introduction

In a nutshell, this work addresses one of the active learning approaches to decrease the learning cost of data stream classifier. We propose to employ so-called *query by clustering* into new classifier training. The main contributions of this work are as follows:

- Presentation of new active data stream classifier learning method.
- Experimental evaluation of the discussed approaches on the basis of diverse benchmark datasets.

A. Monreale et al. (Eds.): ECML PKDD 2018 Workshops, CCIS 967, pp. 80–90, 2019.
https://doi.org/10.1007/978-3-030-14880-5_7

The structure of this article is as follows. Firstly, we describe the proposition of a novel active learning algorithm dedicated to the classification task of drifted streaming data stream. Then we focus on experimental evaluation of the proposed approaches. The final conclusions and proposition of the future works are given thereafter.

1.1   Classification

Classification is an important task among ones studied in the machine learning field. It is a supervised learning problem that consists on assigning classes to observations basing on their attributes. This problem has been a field of study for quite long time. Nevertheless, application of classification in some areas can be discussed. In contemporary world data is generated continuously and cannot be treated as classical classification problems where whole data set is known. Classification systems should create models that are able to adapt to state of the streams which can change over time.

1.2   Data Streams

From the statistical point of view data streams can be described as stochastic processes. Important remark is that such processes are continuous and events are independent [2]. Because of the nature of data streams, one has to keep in mind that there is need to use different approach than in traditional data sets. Samples are arriving on-line and they potentially do not have specified size, so there is need to incrementally update built model with keeping model updated. Such mechanisms for keeping model up to date can be, for instance, windowing or forgetting techniques [2,3]. Another important remark is data can arrive at different velocity – when stream is generating data at high velocity, there is need to use sampling algorithms [4].

Keeping fresh model is important task but there can appear sudden changes in data stream, called *concept drifts*. When such event occur, there is need to rebuild model, as existing one exhibited behavior of data in the past and is no longer applicable.

1.3   Concept Drift

Aforesaid behavior when changes in the data appear is called a *concept drift* [2,5]. This can lead to situation when existing model is no longer relevant, sometimes existing model can be partially accurate. There are many real problems which faces such behavior, such as sensor network analysis [6], fraud detection [7], news categorization [7] or spam detection [7].

There are different methods of classifying concept drift. One can differentiate concept drifts basing on velocity of change: sudden and gradual [5]. Another taxonomy distinguishes two types of drifts which have impact on posterior probabilities: real and virtual concept drifts.

1.4 Labeling

In usual classification problems, cost of labeling is not considered. Typically labels are obtained from the oracle, for instance, human expert. Data streams can have different velocities – when data is arriving too fast, there can be a situation when expert cannot keep pace and some samples need to be discarded. Thus, there is need to implement system that can recognize which incoming samples can be omitted from labeling [8]. Different strategies can be employed to achieve such task, eg. query synthesis or selective sampling [1].

2   Related Works

There are many active learning algorithms. One example can be ACLStream presented in [9]. It is clustering-based approach where clustering is evaluated on every chunk of incoming data. Learning procedure is divided in two steps – *Macro* and *Micro* Steps. First one ranks clusters and the latter ranks instances inside clusters in order to extract the most representative instances for labeling. After labeling procedure clusters are discarded. This algorithm uses fixed number of clusters.

Another algorithm is MINAS [10]. Classes are represented by the micro-clusters which can be incrementally updated. This algorithm has two phases – initial training which uses supervised learning to build a decision model. Second phase consists of online learning using current decision model. Unknown examples are held in short-term memory and when there is sufficient number of examples, they are clustered, creating new micro-clusters.

Applying clustering technique is similar to our approach, however we aim at using any clustering algorithm (by parameterizing it) and using incrementally updated model to extract the most representative samples.

3   Method

The concept of the ALCC algorithm introduces an active learning approach for regular classification algorithms. That approach bases on using clustering algorithms to initially process incoming samples. It is worth mentioning that clustering algorithm Λ must be able to train incrementally. Samples from the data stream are coming in data chunks DS_i, where i is the chunk index. The clustering algorithm Λ is trained incrementally with such chunk. After processing whole chunk clusters are extracted from the evaluated model. Then samples from the chunk are assigned to each cluster, because extracted clusters have only summaries consisting of gravity center and radius. Some samples can be outside any clustering because of the dissimilarities to other samples from the chunk. After assigning task cluster weights can be computed. This computation bases on average distance between points in every cluster. The idea was inspired by the point connectivity measure presented in [11]. After computing distances, they are normalized. For each cluster samples are randomly selected according to the cluster weight and the budget b (eg. 10% randomly selected samples from each cluster) and asked for labels from the oracle. Then the classification algorithm Ψ is trained with such selection.

Presented algorithm is parameterizable and has following parameters:

- chunk length n,
- budget of samples to learn b,
- clustering algorithm Λ,
- option to evaluate points outside of the clusterings p,
- option to evaluate cluster weights d,
- classifying algorithm Ψ.

The idea of the algorithm is presented in Algorithm 1.

Algorithm 1. Active Learning Clustering-Based Algorithm (ALCC)

Require: input data stream, data chunk size, clustering training procedure, classifier
    training procedure, learning budget b, option to evaluate points outside clusterings
1: **repeat**
2:    collect new data chunk DS_i
3:    train incrementally clusterer Λ on the basis of DS_i
4:    extract clustering C from clusterer Λ
5:    fit samples from data chunk DS_i to j clusters from clustering C
6:    **if** compute cluster weights **then**
7:        compute distances between each sample in each cluster from clustering C
8:        normalize distances of samples in clusters
9:        assign weight to each cluster
10:    **else**
11:        assign same weight to each cluster (100%)
12:    **end if**
13:    **for** $k = 1$ **to** j **do**
14:        select randomly b% samples from cluster C_k to $DS'_{i,k}$
15:        ask oracle for labels from $DS'_{i,k}$
16:        train classifier Ψ on the basis $DS'_{i,k}$
17:    **end for**
18:    **if** evaluate points outside clusters **then**
19:        select randomly b% samples not fitting to any cluster to $DS'_{i,(j+1)}$
20:        ask oracle for labels from $DS'_{i,(j+1)}$
21:        train classifier Ψ on the basis $DS'_{i,(j+1)}$
22:    **end if**
23: **until** end of the input data stream

---

4 Experiments

In this section, we describe the details of the experimental study used to verify the usefulness of the proposed methods. The following subsections present goals, used benchmark data streams, experimental set-up, as well as a discussion of obtained results.

4.1 Objective

The main goal of experimental evaluation is to verify the impact of given budget and clustering algorithm on classification accuracy, measured for each processed chunk. Naive Bayes was selected as classifying algorithm. Classifier implementation was done using MOA framework [12] which is incorporated in the Java programming language. Source code of implementation along with experimental results are available on-line in the article repository[1].

4.2 Benchmark Data Streams

Unfortunately, there are not so many benchmark data streams, which may be interpreted as non-stationary ones. We decided to use both real-life data streams and artificially generated ones. Their details have been presented in Table 1.

It is worth mentioning that Zliobaite presents in [17] the problem of autocorrelated data. In presented paper there is dataset which is known to be temporally autocorrelated – Electricity (*elecNormNew*).

To evaluate the proposed methods we to employ *test and train* framework [12], i.e., every classifier is trained on a recent data, but its evaluation (i.e., error estimation and training time) is done on the basis of the following one.

4.3 Algorithm Parameter Setup

For all datasets, classifier presented in this paper was evaluated with different sets of parameters:

- *budget* – from 10% to 80%, with step of 10%,
- *option to compute distances between points in clusters* – on and off,
- *option to compute samples outside of clusterings* – on and off,
- *clustering algorithm* – clustering algorithm to perform initial data analysis – Clustream [18], ClusTree [19], Dstream [20].

Such combinations created set of 96 different parameter groups.

4.4 Results

Due to the stochastic nature of examined algorithms, all experiments were repeated 5 times and averaged results have been presented.

Statistical analysis was done in KEEL software [21]. Friedman $N \times N$ test was used, with Shaffer post-hoc method. Below are presented results for two groups: with and without computing distances between points in clusters. Only top 5 of both algorithm groups are presented in this paper. One can see that variants which used Dstream do not appear in presented results, as they did not outperform the best parameter groups. More results are available in paper's repository, mentioned in Sect. 4.1.

---

[1] https://github.com/jagub2/MOA-ALCClassifier.

Table 1. Data streams used for evaluation

Type	Name	Attributes	Classes	Task	Source
Real	covtypeNorm	54	7	Predict forest cover type from cartographics attributes	[13,14]
	elecNormNew	6	2	Predict change of electricity price in New South Wales, Australia	[13,15]
	poker-lsn	10	10	Predict poker hand	[13]
	sensor	6	54	Predict sensor ID number based on the sensor readings	[16]
Artificial	HyperplaneSlow	10	4	Predict class of rotating hyperplane (slower change)	[12]
	HyperplaneFaster	10	4	Predict class of rotating hyperplane (faster change)	[12]
	LED	24	10	Predicting the digit displayed on a LED segment display (with concept drift)	[12]
	LEDNoDrift	24	10	Predicting the digit displayed on a LED segment display (without concept drift)	[12]
	RandomTree-Recurring	10	4	Stream with randomly generated tree	[12]
	RandomTree-RecurringFaster	10	6	Stream with randomly generated tree	[12]
	SEASudden	3	2	Stream basing on SEA functions, with concept drift	[12]
	SEASudden-Faster	3	2	Stream basing on SEA functions, with concept drift	[12]

Below are presented results of the Friedman test where last column exhibits rank, in Tables 2 and 3 (Table 4).

Results of the Shaffer post-hoc test between presented algorithm and baseline method are depicted in Table 5.

Table 2. Ranked tests of presented algorithm, along with parameter groups. Variant with computing distances between points in clusters.

Algorithm	b	d	p	Clustering algorithm	Rank
ALCC[a]	0.8	✓	✓	Clustream	12.29
ALCC[b]	0.7	✓	✗	Clustream	12.46
ALCC[c]	0.8	✓	✗	Clustream	12.58
ALCC[d]	0.7	✓	✓	ClusTree	13.58
ALCC[e]	0.6	✓	✗	Clustream	13.71
NaiveBayes	—	—	—	—	2.92

Table 3. Ranked tests of presented algorithm, along with parameter groups. Variant without computing distances between points in clusters.

Algorithm	b	d	p	Clustering algorithm	Rank
ALCC[f]	0.8	✗	✗	ClusTree	8.29
ALCC[g]	0.7	✗	✓	ClusTree	8.63
ALCC[h]	0.8	✗	✓	Clustream	9.21
ALCC[i]	0.8	✗	✗	Clustream	9.79
ALCC[j]	0.6	✗	✗	Clustream	9.92
NaiveBayes	—	—	—	—	11.00

Table 4. Presentation of the results for the best combinations of parameters

Data stream	Clusterer	b	d	p	Accuracy	Difference
covtypeNorm	ClusTree	0.8	✗	✗	60.09	0.02
elecNormNew	Clustream	0.8	✗	✗	71.43	0.35
poker-lsn	Clustream	0.8	✗	✗	58.82	0.01
sensor	Clustream	0.6	✗	✗	7.65	0.04
HyperplaneFaster	Clustream	0.6	✗	✗	78.09	0.05
HyperplaneSlow	Clustream	0.6	✗	✗	85.74	−0.13
LED	Clustream	0.6	✗	✗	51.33	0.02
LEDNoDrift	Clustream	0.6	✗	✗	51.43	−0.02
RandomTreeRecurring	ClusTree	0.8	✗	✗	42.31	0.05
RandomTreeRecurringFaster	Clustream	0.8	✗	✓	33.60	0.03
SEASudden	Clustream	0.6	✓	✗	83.68	0.08
SEASuddenFaster	Clustream	0.6	✗	✗	84.51	0.02

Table 5. Shaffer post-hoc comparison between the ALCC and baseline algorithm.

Test	p-value	Outperforms original algorithm?
ALCC[a] vs NB	9.31	✗
ALCC[b] vs NB	8.34	✗
ALCC[c] vs NB	7.68	✗
ALCC[d] vs NB	2.11	✗
ALCC[e] vs NB	3.72	✗
ALCC[f] vs NB	63.63	✓
ALCC[g] vs NB	63.63	✓
ALCC[h] vs NB	63.63	✓
ALCC[i] vs NB	63.63	✓
ALCC[j] vs NB	63.63	✓

4.5 Analysis

For every dataset, one can see that evaluation is very similar to the baseline algorithm. However, presented algorithm uses less labeled data to train the model.

Unfortunately, statistical analysis yields disturbing conclusions. For variant without computing lengths between points in clusters, statistically significant differences are not present in contrast to the baseline algorithm.

For the latter, there are statistically significant differences. However, ranked results present that this variant of algorithm is not outperforming the baseline algorithm.

No statistically significant differences do not necessarily mean bad results. Using less labeled data to train the classifier, as specified by budget to label instances, can produce very similar classifier as the baseline one (Fig. 1).

4.6 Lessons Learnt

To sum up, few observations can be drawn:

- Employing active learning techniques can conduct to create model, similar to the one created by learning all incoming samples, which would maintain similar accuracy but with reduction of number of labeled instances.
- Data from stream can be pre-processed in many ways, in this paper *query by clustering* method was used. Created clusters can be used to group samples. However, this is not the same case as labeling data. High dense clusters have potentially similar samples which could have same label. Less dense clusters, with lesser distances between points in cluster can potentially have samples of different classes.

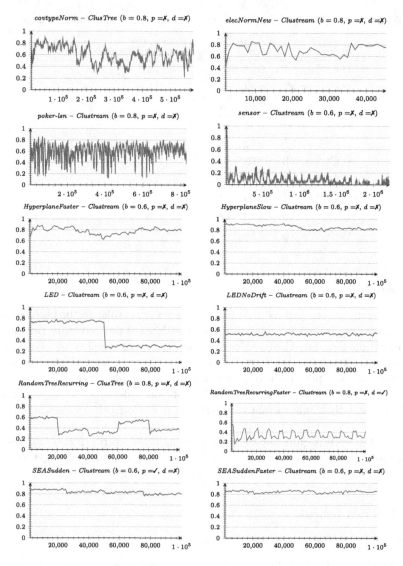

Fig. 1. Stream evaluations for the best set of parameters for ALCC algorithm (red color) versus Naive Bayes evaluation (green color). (Color figure online)

5   Conclusions

The novel active learning classifying algorithm has been proposed. The main idea of pre-processing data is to use clustering algorithm which would gather similar objects in groups. Of course, one has to keep in mind that such groups is not the same thing as labeling data. The results of the experiment show that using less labeled instances to train the classifier can yield very similar model as using data from the whole data stream. However, considering cluster constraints, such as distance between points in clusters in this case, did not exhibit improvement in terms of number of asked labels while maintaining similar accuracy to the baseline classifier.

Presented algorithm is open for modifications, in the future research we will focus on:

- Developing method which will automatically adjust budget, according to the change in accuracy of the classifier,
- Employ more sophisticated method to detect incoming concept drifts.

Acknowledgment. This work is supported the statutory funds of the Department of Systems and Computer Networks, Faculty of Electronics, Wrocław University of Science and Technology.

References

1. Settles, B.: Active Learning. Synthesis Lectures on Artificial Intelligence and Machine Learning, vol. 6, no. 1, pp. 1–114 (2012)
2. Gama, J.: Knowledge Discovery from Data Streams, 1st edn. Chapman & Hall/CRC, Boca Raton (2010)
3. Gama, J., Zliobaite, I., Bifet, A., Pechenizkiy, M., Bouchachia, A.: A survey on concept drift adaptation. ACM Comput. Surv. **46**(4), 44:1–44:37 (2014). http://doi.acm.org/10.1145/2523813
4. Domingos, P., Hulten, G.: Mining high-speed data streams, pp. 71–80. ACM Press (2000)
5. Tsymbal, A.: The problem of concept drift: definitions and related work. Technical report, Trinity College Dublin (2004)
6. Gama, J., Gaber, M.: Learning from Data Streams: Processing Techniques in Sensor Networks. Springer, Heidelberg (2007). https://doi.org/10.1007/3-540-73679-4
7. Zliobaite, I., Pechenizkiy, M., Gama, J.: An overview of concept drift applications. In: Japkowicz, N., Stefanowski, J. (eds.) Big Data Analysis: New Algorithms for a New Society, pp. 91–114. Springer, Cham (2016). https://doi.org/10.1007/978-3-319-26989-4_4
8. Gaber, M.M., Zaslavsky, A., Krishnaswamy, S.: Mining data streams: a review. SIGMOD Rec. **34**(2), 18–26 (2005). http://doi.acm.org/10.1145/1083784.1083789
9. Ienco, D., Bifet, A., Žliobaitė, I., Pfahringer, B.: Clustering based active learning for evolving data streams. In: Fürnkranz, J., Hüllermeier, E., Higuchi, T. (eds.) DS 2013. LNCS (LNAI), vol. 8140, pp. 79–93. Springer, Heidelberg (2013). https://doi.org/10.1007/978-3-642-40897-7_6
10. de Faria, E.R., de Leon Ferreira Carvalho, A.C.P., Gama, J.: MINAS: multiclass learning algorithm for novelty detection in data streams. Data Min. Knowl. Discov. **30**(3), 640–680 (2016). https://doi.org/10.1007/s10618-015-0433-y
11. Kremer, H., et al.: An effective evaluation measure for clustering on evolving data streams. In: Proceedings of the 17th ACM SIGKDD International Conference on Knowledge Discovery and Data Mining, KDD 2011, pp. 868–876. ACM, New York (2011). http://doi.acm.org/10.1145/2020408.2020555

12. Bifet, A., Holmes, G., Kirkby, R., Pfahringer, B.: MOA: massive online analysis. J. Mach. Learn. Res. **11**, 1601–1604 (2010). http://portal.acm.org/citation.cfm?id=1859903

13. Dheeru, D., Taniskidou, E.K.: UCI machine learning repository (2017). http://archive.ics.uci.edu/ml

14. Blackard, J.A., Dean, D.J.: Comparative accuracies of artificial neural networks and discriminant analysis in predicting forest cover types from cartographic variables. Comput. Electron. Agricult. **24**, 131–151 (1999)

15. Harries, M., Wales, N.S.: Splice-2 comparative evaluation: electricity pricing. Technical report (1999)

16. Zhu, X.H.: Stream data mining repository (2010). http://www.cse.fau.edu/~xqzhu/stream.html

17. Zliobaite, I.: How good is the electricity benchmark for evaluating concept drift adaptation. CoRR, abs/1301.3524 (2013). http://arxiv.org/abs/1301.3524

18. Aggarwal, C.C., Han, J., Wang, J., Yu, P.S.: A framework for clustering evolving data streams. In: Proceedings of the 29th International Conference on Very Large Data Bases, VLDB 2003, vol. 29, pp. 81–92. VLDB Endowment (2003). http://dl.acm.org/citation.cfm?id=1315451.1315460

19. Kranen, P., Assent, I., Baldauf, C., Seidl, T.: The ClusTree: indexing micro-clusters for anytime stream mining. Knowl. Inf. Syst. **29**(2), 249–272 (2011)

20. Chen, Y., Tu, L.: Density-based clustering for real-time stream data. In: Proceedings of the 13th ACM SIGKDD International Conference on Knowledge Discovery and Data Mining, KDD 2007, pp. 133–142. ACM, New York (2007). http://doi.acm.org/10.1145/1281192.1281210

21. Alcalá-Fdez, J., Fernández, A., Luengo, J., Derrac, J., García, S.: Keel data-mining software tool: data set repository, integration of algorithms and experimental analysis framework. Multiple-Valued Log. Soft Comput. **17**(2–3), 255–287 (2011). http://dblp.uni-trier.de/db/journals/mvl/mvl17.html#Alcala-FdezFLDG11

Self Hyper-parameter Tuning for Stream Recommendation Algorithms

Bruno Veloso[1,2], João Gama[1,3], Benedita Malheiro[4,5(✉)],
and João Vinagre[1,6]

[1] LIAAD, INESC TEC, Porto, Portugal
{bmveloso,jgama,jnsilva}@inesctec.pt
[2] Research on Economics, Management and Information Technologies - REMIT,
Univ Portucalense, Porto, Portugal
[3] FEP, University of Porto, Porto, Portugal
[4] ISEP, Polytechnic of Porto, Porto, Portugal
mbm@isep.ipp.pt
[5] CRAS, INESC TEC, Porto, Portugal
[6] FCUP, University of Porto, Porto, Portugal

Abstract. E-commerce platforms explore the interaction between users
and digital content – user generated streams of events – to build and
maintain dynamic user preference models which are used to make meaningful recommendations. However, the accuracy of these incremental
models is critically affected by the choice of hyper-parameters. So far, the
incremental recommendation algorithms used to process data streams
rely on human expertise for hyper-parameter tuning. In this work we
apply our Self Hyper-Parameter Tuning (SPT) algorithm to incremental recommendation algorithms. SPT adapts the Melder-Mead optimisation algorithm to perform hyper-parameter tuning. First, it creates
three models with random hyper-parameter values and, then, at dynamic
size intervals, assesses and applies the Melder-Mead operators to update
their hyper-parameters until the models converge. The main contribution of this work is the adaptation of the SPT method to incremental
matrix factorisation recommendation algorithms. The proposed method
was evaluated with well-known recommendation data sets. The results
show that SPT systematically improves data stream recommendations.

Keywords: Parameter tuning · Hyper-parameters · Optimisation ·
Nelder-Mead · Recommendation

1 Introduction

With the increase of strategic information retained by businesses, the adoption
of machine learning algorithms is essential to retrieve valuable information and
increase profits. However, these machine learning tools still face a set of complex
problems such as on-line hyper-parameter optimisation and model selection.

© Springer Nature Switzerland AG 2019
A. Monreale et al. (Eds.): ECML PKDD 2018 Workshops, CCIS 967, pp. 91–102, 2019.
https://doi.org/10.1007/978-3-030-14880-5_8

The hyper-parameter optimisation problem has been addressed in the literature using grid-search [11], random-search [1] and gradient descent [18] algorithms. So far, these approaches have been applied to off-line scenarios since they require train and validation stages. To overcome this limitation, this work focus on the on-line hyper-parameter optimisation for stream-based recommendation.

Self Parameter Tuning (SPT) is a direct-search hyper-parameter optimisation algorithm based on the Nelder-Mead algorithm [22] and dynamic data stream samples. Our proposal applies SPT [27] to stream-based recommendation, continuously searching for the optimal learning rate and regularisation parameter, *i.e.*, for the best incremental matrix factorisation model.

The contribution of this paper is the application of the SPT algorithm to incremental recommendation algorithms. Our extension of the Nelder-Mead algorithm not only processes successfully recommendation problems, but is, to the best of our knowledge, the single one which effectively works with data streams.

The rest of the paper is organised as follows: Sect. 2 describes the related automatic machine learning work; Sect. 3 presents the proposed solution for the identified problem; Sect. 4 describes the experiments and discusses the results obtained; and Sect. 5 presents the conclusions and suggests future developments.

2   Related Work

In machine learning, the ability to select appropriate features, work flows, machine learning paradigms, algorithms, and their hyper-parameters requires expert knowledge [12]. The few contributions found in the literature addressing this progressive automation of machine learning or auto-ML include tools [1,8,26], model selection algorithms [5,6], hyper-parameter optimisation algorithms [9,17,23] and Nelder-Mead optimisation solutions [7,15,24].

These on-line auto-ML tools adopt Bayesian optimisers to tune the hyper-parameters of the specified model [1,8,26]; [1,26] use cross-validation to guide the search direction; and [8] takes into account the performance on similar data sets to improve the efficiency of the algorithm. In terms of automatic hyper-parameter selection, there are several different techniques: (*i*) particle swarm optimisation [6], which is flexible and can be applied to ensemble models [5]; (*ii*) grid search [17], which minimises the estimated error until converges on a local minima; (*iii*) gradient-based search, *e.g.*, Stochastic Gradient Descent (SGD), which converges to an optimal solution [23]; and (*iv*) Nelder-Mead direct search, which relies on heuristics to optimise model parameters [15] or tensor based models [7]. The Nelder-Mead algorithm has been used together with exponentially decay centrifugal forces to improve the results at the cost of the number of iterations needed to converge [14] as well as with reinforcement techniques (e-greedy) to select the best model of each iteration [24].

Our proposal differs from all the above because it operates on-line by automatically adjusting the hyper-parameters of the models based on the stream of events. Nevertheless, it is also applicable to off-line batch learning.

3 Self Parameter Tuning

This paper presents the application of the SPT algorithm to stream-based recommendation algorithms. The SPT algorithm was designed to optimise a set of hyper-parameters,namely the learning rate and regularisation parameter. We adopt a direct-search algorithm, that uses heuristics to avoid algorithms which rely on hyper-parameters. Specifically, we adapt the Nelder-Mead method [22] to work with stream-based recommendation algorithms. Fig. 1 represents the

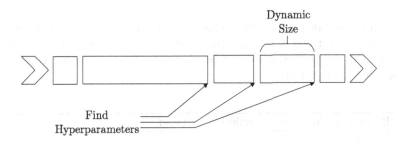

Fig. 1. Application of the proposed algorithm to the data stream.

application of the proposed algorithm. In particular, to find a solution for n hyper-parameters, it requires $n + 1$ input models, *e.g.*, to optimise two hyper-parameters, the algorithm needs three alternative input models, corresponding to the three vertexes of the Nelder-Mead algorithm. The models are initialised with randomly selected learning rate and regularisation parameters, and the Nelder-Mead operators are then applied over dynamic sample intervals. The algorithm processes each data stream sample, using the three models until they converge. The following subsections describe the implemented Nelder-Mead algorithm, including the dynamic sample size selection.

3.1 Nelder-Mead Optimisation

This algorithm is a simplex search algorithm for multidimensional unconstrained optimisation without derivatives. The vertexes of the simplex, which define a convex hull shape, are iteratively updated in order to sequentially discard the vertex associated with the largest cost function value.

The Nelder-Mead algorithm applies four simple operations to the three vertexes (models): *reflection*, *shrinkage*, *expansion* and *contraction* (Fig. 2). The three vertexes are ordered by root mean square error (RMSE) value: best (B), good (G), which is the second best, and worst (W).

While Algorithm 1 implements the reflection and extension operations, Algorithm 2 addresses the contraction and shrinkage operations. Each operation computes an additional set of vertexes (midpoint M, reflection R, expansion E, contraction C and shrinkage S) and ensures they belong to the search space.

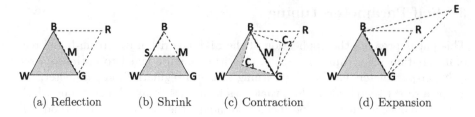

(a) Reflection	(b) Shrink	(c) Contraction	(d) Expansion

Fig. 2. Nelder-Mead operations

First, Algorithm 1 determines the midpoint (M) of the best side of the triangle – connecting the best vertex (B) and the good vertex G – as well as the reflection point (R). After this initial step, it heuristically decides whether to reflect or expand (lines 3, 4 and 8).

Algorithm 1. Nelder-Mead - reflect or expand

1: $M = (B + G)/2$
2: $R = 2M - W$
3: **if** $f(R) < f(G)$ **then**
4:     **if** $f(B) < f(R)$ **then**
5:         $W = R$
6:     **else**
7:         $E = 2R - M$
8:         **if** $f(E) < f(B)$ **then**
9:             $W = E$
10:        **else**
11:            $W = R$
12:        **end if**
13:    **end if**
14: **end if**

Algorithm 2 calculates the contraction point (C) of the worst side of the triangle – the midpoint between the worst vertex (W) and the midpoint M – and shrinkage point (S) – the midpoint between the best (B) and the worst (W) vertexes. Then, it determines whether to contract or shrink based on the set of predetermined heuristics (lines 3, 4, 8, 12 and 15).

In this case, we intend to optimise the learning rate and the regularisation parameter, which are constrained to values between 0 and 1. The violation of this constraint results in the adoption of the nearest lower or upper bound.

3.2   Adaptive Sample Size

The outcome of the Nelder-Mead algorithm depends on the sample size. We calculate a dynamic sample size based on the RMSE metric every time the Nelder-Mead tries to find an optimal solution. The sample size S_{size} is given by Eq. 1 where σ represents the RMSE standard deviation and M is the desired error margin. We use 95 % in our experimental work.

Algorithm 2. Nelder-Mead - contract or shrink

1: $M = (B + G)/2$
2: $R = 2M - W$
3: **if** $f(R) \geq f(G)$ **then**
4:      **if** $f(R) < f(W)$ **then**
5:          $W = R$
6:      **else**
7:          $C = (W + M)/2$
8:          **if** $f(C) < f(W)$ **then**
9:              $W = C$
10:         **else**
11:             $S = (B + W)/2$
12:             **if** $f(S) < f(W)$ **then**
13:                 $W = S$
14:             **end if**
15:             **if** $f(M) < f(G)$ **then**
16:                 $G = M$
17:             **end if**
18:         **end if**
19:     **end if**
20: **end if**

$$S_{size} = \frac{4\sigma^2}{M^2} \qquad (1)$$

However, to avoid using small samples, we defined a lower bound of 30 samples.

4   Experimental Evaluation

The following subsections describe the experiments performed, including the data sets, the evaluation metrics and protocol, the tests and the results. The experiments were performed with an Intel Xeon CPU E5-2680 2.40 GHz Central Processing Unit (CPU), 32 GiB DDR3 Random Access Memory (RAM) and 1 TiB of hard drive platform running the Ubuntu 16.04. The SPT approach was compared against a default hyper-parameter initialisation – hereafter called baseline. The baseline hyper-parameter initialisation was, 1.0 for the learning rate and 0.05 for the regularisation parameters.

4.1   Data Sets

For the experiments, we selected the following recommendation data sets: (i) MovieLens 100k (ML100k) [20] contains information about 943 users and 1682 movies, including 100 000 user ratings together with time stamps; (ii) MovieLens 1M (ML1M) [21] holds information about 6040 users and 3900 movies, including 1 000 209 user ratings together with time stamps; (iii) Jester [2] data set stores information about 59 132 users and 150 jokes, including 1.7 million user ratings; and (iv) GoodBooks [13] data set contains information on 10 000 books, including 1 000 000 user ratings.

4.2   Evaluation Metrics and Protocol

The evaluation protocol defines the data ordering, partitioning, distribution and evaluation metrics. To evaluate the proposed method we applied two different protocols: holdout evaluation [16] and the predictive sequential (prequential) evaluation [10]. The holdout evaluation protocol is used to find an optimal solution for the hyper-parameters and verify the reproducibility of the algorithm. Then, we apply the prequential evaluation to the data as a stream to assess the performance of our method.

In terms of evaluation metrics we adopt the incremental RMSE adopted by Takács *et al.* (2009) [25], which is calculated incrementally after each new viewer rating event. Additionally, we calculate incrementally the Recall@N proposed by Cremonesi *et al.* [3]. For each new event, we randomly select 1000 items not yet rated by the active user, add the newly rated item and, then, make predictions for this subset of 1001 items. Finally, we sort these 1001 items by descending prediction value and, if the newly rated item belongs to the list of the top N viewer predicted items, we count a hit.

Figure 3 illustrates the holdout evaluation where the entire data stream is ordered temporally and, then, partitioned in two halves: 50 % to "Train" and the remaining 50 % to "Test". First, the holdout algorithm finds an optimal solution for the selected hyper-parameters using the train data. Then, it builds a model using the train data and the identified optimal hyper-parameters. Finally, the holdout algorithm updates and evaluates the created model using the test data. The holdout protocol was repeated 30 times to compute the average and standard deviation of the evaluation metrics.

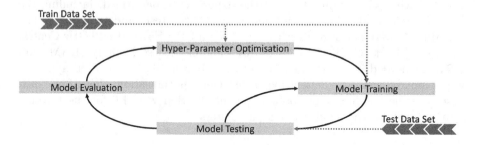

Fig. 3. Holdout – data splitting and processing

In the case of the prequential protocol, the data stream is dynamically partitioned into subsets which are simultaneously used for training and testing as represented in Fig. 4. First, the data are ordered temporally, then, they are used to build incrementally the three models and, finally, the results are evaluated with a sliding window of 1000 instances. In order to produce the best recommendations, the prediction model used throughout the experiment is dynamic. In fact, this dynamic model corresponds to the best model found so far by the SPT algorithm.

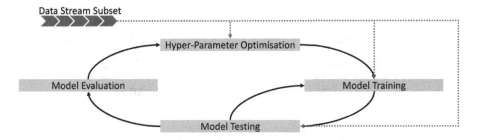

Fig. 4. Prequential – data splitting and processing

4.3 Significance Tests

To detect the statistical differences between the proposed and the baseline approaches we applied three different significance tests: (i) the Wilcoxon test [28] to verify if the mean ranks of two samples differ; (ii) the McNemar test [19] to assess if a statistically significant change occurs on a dichotomous trait at two time points on the same population; and (iii) the critical distance measure proposed by [4] for a graphical interpretation of the statistical results. We define a 5 % of significance level for all tests. The goal of the Wilcoxon and McNemar tests is to reject the null-hypothesis, *i.e.*, that both approaches have the same performance. We run 30 trials for each experiment. At 5 % significance, the critical value of McNemar test (MT_{crit}) is 3.84 and the critical value of the Wilcoxon test (WT_{crit}) is 137. In the case of the McNemar, two samples are statistically different if $MT_{stat} > MT_{crit}$, whereas, in the case of Wilcoxon, two samples are statistically different if the $|WT_{stat}| > WT_{crit}$.

4.4 Experiments

The goal is to optimise the learning rate and regularisation hyper-parameters of the recommendation algorithm proposed by Takács *et al.* (2009) [25]. First, we created three identical initial models with randomly selected learning rate and regularisation values and, then, applied our hyper-parameter optimisation algorithm. Figure 5 shows that the convergence of the three recommendation models occurs in less than 5000 events with all data sets.

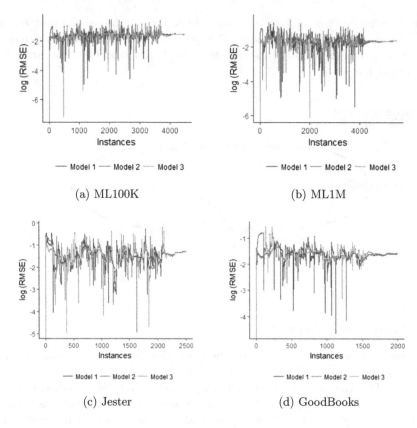

Fig. 5. Recommendation - model convergence

After the verification of the model convergence, we applied the holdout evaluation protocol to assess the performance of the algorithm with the new hyperparameters. The experiment was performed 30 times to compute the average and standard deviation of the RMSE and Recall@10 for the SPT and baseline (B) approaches. Table 1 not only displays these results, but highlights for each data set the best case of each evaluation metric, including the corresponding coefficients of variation (CV). The ML100k data set displays a RMSE decrease of 1.4 % and a Recall@10 increase of 2.1 %. With ML1M, the prediction error decreases 1.9 % and the Recall@10 drops 1.9 %. The Jester data set shows an improvement of 5.5 % and 6.6 % in terms of RMSE and Recall@10, respectively. Finally, the GoodBooks data set presents a decrease of 2.8 % in RMSE and an increase of 1.0 % in Recall@10. Regarding the holdout results, the statistical results for the Wilcoxon and McNemar tests reject the null hypothesis, regardless of the data set. The McNemar test statistic value (MT_{stat}) is 28.03 which corresponds to a p-value of 1.19×10^{-7} and the Wilcoxon statistic value (WT_{stat}) is 465 with a p-value of 1.86×10^{-9}. The calculated McNemar and Wilcoxon test statistic values are higher than the corresponding reference values for p-value $= 0.05$ and

Table 1. Recommendation – holdout results

Dataset	Approach	Metric	μ	CV (%)
ML100K	B	RMSE	2.046×10^{-1}	0.074
		Recall@10	0.097×10^{-1}	3.166
	SPT	RMSE	2.018×10^{-1}	0.066
		Recall@10	0.099×10^{-1}	3.427
ML1M	B	RMSE	2.016×10^{-1}	0.048
		Recall@10	0.106×10^{-1}	1.908
	SPT	RMSE	1.978×10^{-1}	0.038
		Recall@10	0.104×10^{-1}	1.672
Jester	B	RMSE	2.400×10^{-1}	0.226
		Recall@10	0.855×10^{-1}	0.265
	SPT	RMSE	2.269×10^{-1}	0.192
		Recall@10	0.911×10^{-1}	0.284
GoodBooks	B	RMSE	1.952×10^{-1}	0.013
		Recall@10	0.988×10^{-1}	0.325
	SPT	RMSE	1.897×10^{-1}	0.023
		Recall@10	0.996×10^{-1}	0.449

their significance levels are smaller than 0.05. Fig. 6 plots the critical distance between the proposed and baseline optimisation algorithms, showing that they are statistically different. The critical distance between both approaches was determined using the Nemenyi test.

Fig. 6. Recommendation – holdout Critical Distance

The prequential evaluation shows that the proposed dynamic model outperforms the baseline approach. Figure 7 displays the relative RMSE results between the baseline and SPT methods. Considering the Recall@10, there is an increase of 1.8 % with ML100k, 16.5 % with ML1M, 6.3 % with Jester and a decrease of 19.5 % in the case of GoodBooks.

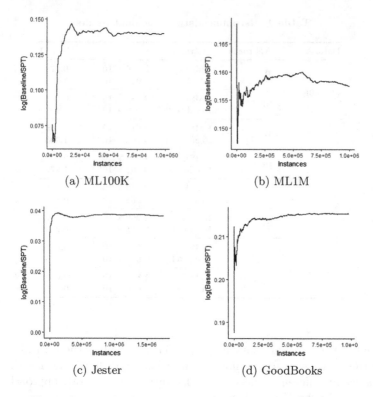

(a) ML100K

(b) ML1M

(c) Jester

(d) GoodBooks

Fig. 7. Recommendation – relative RMSE prequential results

5   Conclusions

This paper describes the application of SPT to incremental recommendation algorithms. In this case, SPT was used to find dynamically the best learning rate and regularisation hyper-parameters.

The main contribution of this paper is an extension of the Nelder-Mead optimisation algorithm to stream-based recommendation. The SPT algorithm is, in terms of existing hyper-parameter optimisation algorithms, less computationally expensive than Bayesian optimisers, stochastic gradients or even grid search algorithms. This proposal is, to the best of our knowledge, the single one which effectively works with data streams in a recommendation scenario.

Taking into consideration that the selection of the hyper-parameters has a substantial impact on the outcome of recommendation algorithms, we applied the proposed method and studied its performance in with holdout and prequential evaluation protocols. The results shows that, not only our algorithm converged rapidly with both evaluation protocols, but also outperformed the baseline results.

Future work will includes three key points: (i) application of the algorithm to classification algorithms; (ii) selection of machine learning models; and (iii) thorough comparison with other optimisation algorithms.

Acknowledgements. This research was carried out in the framework of the project TEC4Growth - RL SMILES - Smart, mobile, Intelligent and Large Scale Sensing and analytics NORTE-01-0145-FEDER-000020 which is financed by the north Portugal regional operational program (NORTE 2020), under the Portugal 2020 partnership agreement, and through the European regional development fund.

References

1. Bergstra, J., Bengio, Y.: Random search for hyper-parameter optimization. J. Mach. Learn. Res. **13**(1), 281–305 (2012)
2. Berkeley University: Jester data set. Accessed Mar 2018
3. Cremonesi, P., Koren, Y., Turrin, R.: Performance of recommender algorithms on top-n recommendation tasks. In: Proceedings of the Fourth ACM Conference on Recommender Systems, RecSys 2010, pp. 39–46. ACM, New York (2010)
4. Demšar, J.: Statistical comparisons of classifiers over multiple data sets. J. Mach. Learn. Res. **7**, 1–30 (2006)
5. Escalante, H.J., Montes, M., Sucar, E.: Ensemble particle swarm model selection. In: The 2010 International Joint Conference on Neural Networks (IJCNN), pp. 1–8. IEEE (2010)
6. Escalante, H.J., Montes, M., Sucar, L.E.: Particle swarm model selection. J. Mach. Learn. Res. **10**(Feb), 405–440 (2009)
7. Fernandes, S., Tork, H.F., Gama, J.: The initialization and parameter setting problem in tensor decomposition-based link prediction. In: 2017 IEEE International Conference on Data Science and Advanced Analytics (DSAA), pp. 99–108, October 2017
8. Feurer, M., Klein, A., Eggensperger, K., Springenberg, J., Blum, M., Hutter, F.: Efficient and robust automated machine learning. In: Advances in Neural Information Processing Systems, pp. 2962–2970 (2015)
9. Finn, C., Abbeel, P., Levine, S.: Model-agnostic meta-learning for fast adaptation of deep networks. In: Precup, D., Teh, Y.W. (eds.), Proceedings of the 34th International Conference on Machine Learning, Volume 70 of Proceedings of Machine Learning Research, 06–11 August 2017, International Convention Centre, Sydney, Australia, pp. 1126–1135. PMLR (2017)
10. Gama, J., Sebastião, R., Rodrigues, P.P.: On evaluating stream learning algorithms. Mach. Learn. **90**(3), 317–34 (2013)
11. Hsu, C.-W., Chang, C.-C. and Lin, C.-J., et al.: A practical guide to support vector classification (2003)
12. Hutter, F., et al.: In: AutoML Workshop @ ICML 2014 (2014). Accessed 18 July 2018
13. Kaggle: Goodbooks data set. Accessed Mar 2018
14. Kar, R., Konar, A., Chakraborty, A., Ralescu, A.L., Nagar, A.K.: Extending the Nelder-Mead algorithm for feature selection from brain networks. In: 2016 IEEE Congress on Evolutionary Computation (CEC), pp. 4528–4534. IEEE (2016)
15. Koenigstein, N., Dror, G., Koren, Y.: Yahoo! music recommendations: modeling music ratings with temporal dynamics and item taxonomy. In Proceedings of the Fifth ACM Conference on Recommender Systems, pp. 165–172. ACM (2011)

16. Kohavi, R.: A study of cross-validation and bootstrap for accuracy estimation and model selection. In: Proceedings of the 14th International Joint Conference on Artificial Intelligence - Volume 2, IJCAI 1995, pp. 1137–1143. Morgan Kaufmann Publishers Inc., San Francisco (1995)

17. Kohavi, R., John, G.H.: Automatic parameter selection by minimizing estimated error. In: Machine Learning Proceedings 1995, pp. 304–312. Elsevier (1995)

18. Maclaurin, D., Duvenaud, D., Adams, R.: Gradient-based hyperparameter optimization through reversible learning. In: Proceedings of the 32nd International Conference on International Conference on Machine Learning - Volume 37, ICM 2015, pp. 2113–2122. JMLR.org. (2015)

19. McNemar, Q.: Note on the sampling error of the difference between correlated proportions or percentages. Psychometrika **12**(2), 153–157 (1947)

20. MovieLens: Movielens 100k data set. Accessed Mar 2018

21. MovieLens: Movielens 1M data set. Accessed Mar 2018

22. Nelder, J.A., Mead, R.: A simplex method for function minimization. Comput. J. **7**(4), 308–313 (1965)

23. Nichol, A., Schulman, J.: Reptile: a scalable metalearning algorithm. arXiv e-prints, March 2018

24. Pfaffe, P., Tillmann, M., Walter, S., Tichy, W.F.: Online-autotuning in the presence of algorithmic choice. In: 2017 IEEE International Parallel and Distributed Processing Symposium Workshops (IPDPSW), pp. 1379–1388. IEEE (2017)

25. Takács, G., Pilászy, I., Németh, B., Tikk, D.: Scalable collaborative filtering approaches for large recommender systems. J. Mach. Learn. Res. **10**, 623–656 (2009)

26. Thornton, C., Hutter, F., Hoos, H.H., Leyton-Brown, K.: Auto-weka: combined selection and hyperparameter optimization of classification algorithms. In: Proceedings of the 19th ACM SIGKDD International Conference on Knowledge Discovery and Data Mining, KDD 2013, pp. 847–855. ACM, New York (2013)

27. Veloso, B., Gama, J., Malheiro, B.: Self hyper-parameter tuning for data streams. In: Soldatova, L., Vanschoren, J., Papadopoulos, G., Ceci, M. (eds.) DS 2018. LNCS (LNAI), vol. 11198, pp. 241–255. Springer, Cham (2018)

28. Wilcoxon, F.: Individual comparisons by ranking methods. Biomet. Bull. **1**(6), 80–83 (1945)

Deep Online Storage-Free Learning on Unordered Image Streams

Andrey Besedin[1](✉), Pierre Blanchart[1], Michel Crucianu[2], and Marin Ferecatu[2]

[1] CEA, LIST, Laboratoire d'Analyse de Donnes et Intelligence des Systemes, Digiteo Labs Saclay, 91191 Gif-sur-Yvette Cedex, France
{andrey.besedin,pierre.blanchart}@cea.fr
[2] Centre d'études et de recherche en informatique et communications, Le CNAM, 292 rue Saint-Martin, 75003 Paris, France
{michel.crucianu,marin.ferecatu}@cnam.fr

Abstract. In this work we develop an online deep-learning based approach for classification on data streams. Our approach is able to learn in an incremental way without storing and reusing the historical data (we only store a recent history) while processing each new data sample only once. To make up for the absence of the historical data, we train Generative Adversarial Networks (GANs), which, in recent years have shown their excellent capacity to learn data distributions for image datasets. We test our approach on MNIST and LSUN datasets and demonstrate its ability to adapt to previously unseen data classes or new instances of previously seen classes, while avoiding forgetting of previously learned classes/instances of classes that do not appear anymore in the data stream.

Keywords: Deep learning · GAN · Data streams · Classification

1 Introduction

In recent years methods based on Deep Learning have become state of the art in numerous applications, such as image and signal classification [7], object detection [10] and segmentation [4], natural language processing [2,11] and many others. Despite its popularity and efficiency, most of the currently existing applications are based on offline learning where all the data are constantly available during training. On the other hand, scenarios where data arrive continuously in large quantities and have to be integrated into the learning models in real time are starting to get more and more attention from the Machine Learning community.

In this context, the main problem is that most of the Deep Learning methods are prone to forgetting the concepts that are no longer represented by the dataset they are trained on. In literature this phenomena is known as catastrophic forgetting [8]. The current solution for this problem is to store the whole

A. Monreale et al. (Eds.): ECML PKDD 2018 Workshops, CCIS 967, pp. 103–112, 2019.
https://doi.org/10.1007/978-3-030-14880-5_9

dataset to be able to reuse all the data samples at any moment. At the same time, training Neural Networks is based on gradient backpropagation, which is slow and often requires passing through the dataset many times.

The described problems impose hard constraints for applications which require continuous learning to adapt to changing environments, and, especially, for distributed applications on devices with limited storage and computational resources, like smartphones or small private servers.

In this paper we introduce a method that uses Generative Adversarial Networks [3] to model the real data distribution and replace the necessity of storing and reusing historical data when performing online classification learning on data streams. We test our method on MNIST and LSUN datasets and show that it allows to efficiently train classifiers on multi-class streams of data with possible concept drifts [12] with no need of retraining the model on historical data.

The present proposal is based on our previous work [1] and improves upon it in several ways: first, we extend our framework to handle data coming continuously and in random order, which corresponds to a much more realistic situation; second, we test the framework on the much larger LSUN database [1] with more complex data; third, we quantify the loss (in classification accuracy) when using our method compared to the offline situation. We also study forgetting/classification improvement behavior of our approach on classes that were initially present or appear at some moment of the stream.

The rest of the paper is organized as follows: In Sect. 2 we present our method for online classification of unordered streams without data storage, followed in Sect. 3 by the experimental validation. Section 4 concludes the paper by a discussion of our work and suggests several directions for further development.

2 Proposed Method

The goal of this work is to develop a method allowing to train classifiers on data streams with time changing environment described by the set of the data classes currently present in the stream. The main issue with such a task is that neural network based classifiers tend to forget already learned classes if the corresponding data is removed from the training set, which is a realistic scenario in evolving data streams.

Intuitively, there are two possible ways to handle the problem of forgetting in Neural Networks. From one hand, one could try to control the way the backpropagation works when updating the classifier and to avoid strong modifications of the weighs of the network that are important for correct classification of those classes. From the other hand, forgetting is caused by the absence of corresponding data. Storing and reusing the data itself helps in batch learning, but is hardly feasible in the continuous massive data stream setup. Storing only partial information from missing data or some representation of it should help. In this paper we focus on second idea and propose to train generative models in order to replace the necessity of storing and reusing of historical data.

---

[1] http://lsun.cs.princeton.edu/2017/.

2.1   Replacing Original Data by Generators

To avoid storing historical data we train generative models to learn the distribution of the original data and use them to produce synthetic data samples to replace the original ones. To do so, we use Generative Adversarial Networks that have recently shown excellent ability to learn data distributions on image datasets and generate samples that are very similar to the original images. More specifically, we use a Deep Convolutional version of GAN (DCGAN [9]) in which, comparing to original GAN and its other convolutional modifications, all the pooling layers are replaced by stride convolutions. At the same time, the proposed architecture does not have any fully connected layers, uses Batch Normalization [6] and ReLU activation function are replaced by LeakyReLU in the discriminator network. All the described changes show better stability during training and allow to learn higher resolution models.

2.2   Batch Classification on Generated Data

In our previous work [1] we introduced quantitative metrics to evaluate generative and representative capacities of generative models on a given dataset. We demonstrated that, according to those metrics, DCGANs are able to represent well the original data and to generalize over it, i.e. allow a classifier trained on generated data to have good generalization abilities on unseen test data.

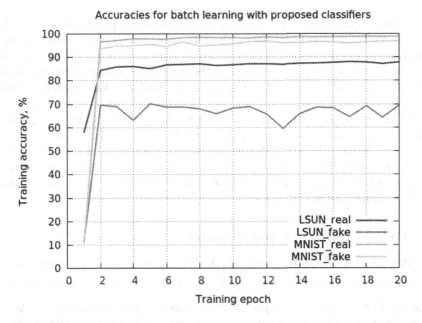

Fig. 1. Batch training accuracy on the original validation data for MNIST and LSUN dataset when trained on real vs. generated data

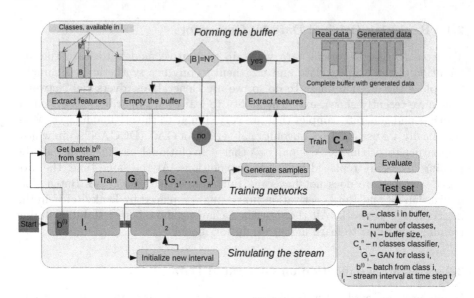

Fig. 2. Stream classification scheme (for LSUN dataset, on MNIST no feature extraction is performed), described in this work. Stream is represented as an infinite sequence of data intervals. Each processing block is described in corresponding subsection of Sect. 2.3. All the green boxes represent the processes, described in details in the paper. (Color figure online)

In order to check if DCGAN-based generators can represent more complex datasets, we train it on LSUN in batch mode, one generator per class. We then train two classifiers: first one on the original data that was used to train the generators, and second one on purely generated data produced by pretrained DCGANs. We then test obtained classifiers on validation set consisting of previously unseen real images and compare the classification accuracies of both. From Fig. 1 we observe that using generated data to train a classifier results in a decrease in classification accuracy (MNIST: $99.14\% \rightarrow 97.16\%$; LSUN: $88.69\% \rightarrow 70.22\%$), especially for LSUN dataset. Nevertheless, we find this decrease an acceptable trade-off to be able to pass to completely online classification training scenario with no necessity to store historical data, especially taking into account the data complexity of LSUN dataset.

2.3 Online Learning on Data Streams

Simulating Data Stream. In this work we consider the case where data arrive continuously in stream. Let $E = \bigcup_{k=1}^{\infty} E_k$ be an environment emitting data continuously in time, where E_k represents the subset of E corresponding to class i. We will make an assumption that data is sampled with the examples of unique type and format (e.g. RGB images of same size, sound recordings of given length, etc.). We will consider that each class, when it appears in the stream, lasts for some period of time and emits at least b samples. We then can say that we receive data from stream in the form of batches of size b, where each batch contains only the elements of one class. Since the datasets we use are static and our goal is to work with data streams, for our experiments we need to define the way data will arrive during training.

We start by assuming that the stream is divided into time intervals. Every interval contains at least two and at most M distinct data classes. Each time a new interval is started we remove several classes from the previous interval and add new classes from E, so that the new interval always contains at least one class from the previous one (to simulate environment continuity) and never exceeds M classes. Every class, when it appears in the stream, emits a random number of batches. The durations of each interval and sub-interval, corresponding to a given class, are taken randomly from corresponding predefined ranges.

Forming the Buffer. Let us also initialize a data buffer B of size $N \times b$, that will serve to collect data from batches in order to use it later to train a classifier. We will fill in the buffer until the number of batches of one of the classes reaches the buffers limit size. After that we complete buffer to have equal number of images for each class by generating samples from all the pretrained generators (Fig. 2, Forming the buffer), and send obtained data to train the classifier. We then empty the buffer and start filling it again.

Network Training. When starting the online training on stream we consider that we already have pretrained generative models for some of the classes from the dataset, as well as the pretrained classifier for those data classes. Each time a new class appears in the stream we initialize a new GAN for it. We train the GANs with batches of corresponding classes directly when they appear in stream. The classification network is trained each time the data buffer is complete and its performance is evaluated on the test set at the end of each stream interval. Figure 2 shows the full schematic representation of the proposed framework.

3 Experimental Results

3.1 Datasets and Data Preparation

To test the hypothesis proposed in this paper we perform our evaluations first on the MNIST dataset, which is usually used as a baseline dataset in many ML-based studies in image analysis, and then on the LSUN dataset to check its performance on more complex data.

In every experimental setup, independently from the dataset, we train one generator per data class.

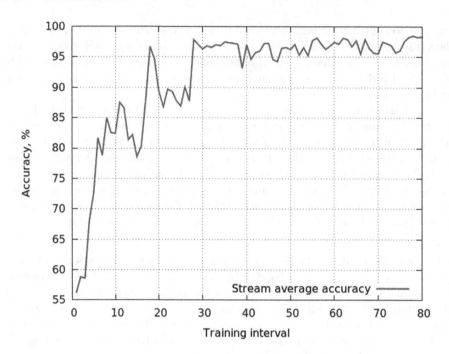

Fig. 3. Classification accuracy during online stream training for MNIST dataset

MNIST is a collection of gray-scale images of hand-written digits of 28×28 pixels each. The images include 10 data classes, each corresponding to a separate number from 0 to 9. The training set includes 6000 images per class and the test set 1000 images per class. No spatial transformation was applied on MNIST for either classification or GAN training. The classification network we use on MNIST consists of two convolutional with max pooling layers (with resp. 16 and 32 feature maps using 4×4 kernels), followed by three fully connected layers (512×512, 512×128 and 128×10) with ReLU activation function except for the output layer.

LSUN is a collection of RGB images of size at least 256×256 pixels each. The dataset includes 10 classes of scenes (bedroom, bridge, church outdoor, classroom, conference room, dining room, kitchen, living room, restaurant and tower), with the smallest class containing around 126k images and the biggest one over 3 millions of images. We extracted 5k images from each class to use them as a validation set for classification, the rest of the images were used to form the stream and train both generative models and classifier.

Every image is transformed to square shape by cutting its sides. DCGAN in its original formulation does not work on big size images, but works perfectly well on images of size 64×64 pixels and less. Also, since our goal was to simulate a data stream with only unique samples, we needed to perform some data augmentation on the dataset. For these reasons, we rescaled LSUN images to the size of 96 × 96 pixels and randomly cropped them to 64 × 64 pixels each time they appear in stream.

Training state-of-the-art classifiers for large complex datasets usually requires very deep network architectures and takes a lot of time and resources. Getting a performance similar to the state-of-the-art batch learning scenario was not the intention of this work, so, we used a few shortcuts that allowed us to speed up training, to the expense of a slight decrease in classification accuracy. More precisely, we rescaled the original images, as well as the generated 64 × 64 pixels images, to a 224 × 224 size, and, passed them through the convolutional layers of the ResNet-200 network ([5]), a convolutional network pretrained on the ImageNet dataset. The latter was thus used as a feature extractor. Four fully-connected layers with ReLU activations (except for the output layer) were added on top of the feature extraction block to form the 10-class classification network $(2048 \times 1024 \rightarrow 1024 \times 512 \rightarrow 512 \times 128 \rightarrow 128 \times 10)$.

3.2   Online Classification

We performed online classifier training on MNIST and LSUN datasets, which were streamed in the way described in Sect. 2.3.

In our online-learning-on-stream scenario, we achieved a maximum accuracy of 98.64% on MNIST (Fig. 3) and 77.59% for LSUN on 10 classes, which is comparable to the results of our batch experiments where the classifiers are trained on only generated data from pretrained DCGANs. Comparing the results of online classification on stream with batch classification allow us to quantify the loss of performance, to the best of our knowledge there exist very few works on online classification on evolving streams of complex data, and no established baseline for evaluation.

The online training on a stream is quite unstable for the LSUN dataset and classification accuracy varies a lot from one interval to the other as can be seen from the accuracy standard deviation that is plotted in light blue on Fig. 4. Still, performing mean filtering of classification accuracy over several successive intervals results in a curve that shows stable progression (the results for LSUN presented on Fig. 4 are averaged/mean-filtered over 80 successive training intervals). To find the reasons for the training instability in the stream scenario, we performed a test in a similar scenario with the only difference that the original images from the stream were used only to train generative models, while the classifier was trained with only generated data at the end of each interval (Fig. 5). The yellow curve on the diagram of Fig. 5 represents the average classification accuracy as a function of the average number of images (×1000) seen by each DCGAN. The red and blue lines represent respectively the average accuracy over pretrained classes and the average accuracy over the classes appearing at some point of the stream. We can see on Fig. 5 how the amount of images fed to the DCGAN improves the classification accuracy: each generative model received about 1M real images before it started to have a positive influence on classification training, while after 5M of images per class we do not see global

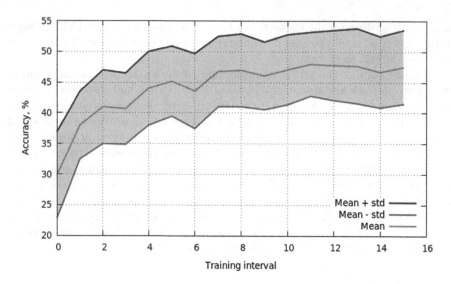

Fig. 4. Classification accuracy during online stream training for LSUN dataset. Each point on the graph corresponds to the average value over 80 training intervals (Color figure online)

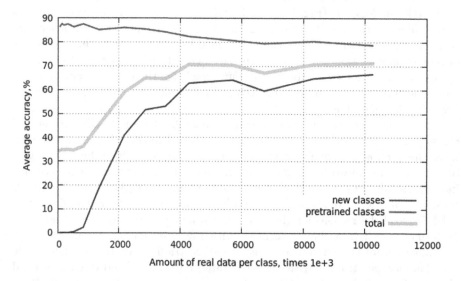

Fig. 5. Average classification accuracy during stream training when the classifier is trained only on generated data. The curves correspond respectively to the average performance over all classes (gold), the average performance over classes pretrained before the beginning of the stream (red), and, the average performance over classes introduced during the stream (blue) (Color figure online)

improvements in classification accuracy. The training in such a scenario appeared to be much more stable than when using a mixture of real and generated data to train the classifier.

We think that the training instability on the LSUN dataset is due to the complexity of the database (for the MNIST dataset, the training is rather stable), but, also, to multiple image rescaling steps rendered necessary by the incapacity of DCGANs to work directly with full-size 256×256 images. The classification accuracy during training might also be limited due to the fact that we use a pretrained network on ImageNet to perform the feature extraction step for the classifier, and that this feature extraction layer is not retrained in our stream scenario.

4   Conclusions

In this work we presented a new method for online classification on data streams. We defined streaming scenarios on the MNIST and LSUN datasets, and validated our online-learning-on-stream method on these datasets by showing that it is able to efficiently learn to classify complex image data from a time-evolving stream, with no need to store historical data. We also showed that DCGAN-based models are able to generate samples representative enough to replace the real data when training a classifier. Our online-learning-on-stream method showed on one side a strong capacity to adapt to unseen data classes appearing at different time of the stream, and, on the other side, did not lead to catastrophic forgetting of previously seen data.

The current approach requires quite a large amount of data per class which are not always available in the case of real data streams. We plan to tackle this problem in a future work by thinking of more reactive and efficient retraining procedures for the classification model (and eventually the generative models), able to retrain a network on a few data without forgetting of previously seen data.

References

1. Besedin, A., Blanchart, P., Crucianu, M., Ferecatu, M.: Evolutive deep models for online learning on data streams with no storage. In: 2nd ECML/PKDD 2017 Workshop on Large-Scale Learning from Data Streams in Evolving Environments (2017)
2. Collobert, R., Weston, J., Bottou, L., Karlen, M., Kavukcuoglu, K., Kuksa, P.: Natural language processing (almost) from scratch. J. Mach. Learn. Res. **12**(Aug), 2493–2537 (2011)
3. Goodfellow, I., et al.: Generative adversarial nets. In: Advances in Neural Information Processing Systems, pp. 2672–2680 (2014)
4. He, K., Gkioxari, G., Dollár, P., Girshick, R.: Mask R-CNN. arXiv preprint arXiv:1703.06870 (2017)
5. He, K., Zhang, X., Ren, S., Sun, J.: Deep residual learning for image recognition. In: Proceedings of the IEEE Conference on Computer Vision and Pattern Recognition, pp. 770–778 (2016)

6. Ioffe, S., Szegedy, C.: Batch normalization: accelerating deep network training by reducing internal covariate shift. arXiv preprint arXiv:1502.03167 (2015)
7. Krizhevsky, A., Sutskever, I., Hinton, G.E.: Imagenet classification with deep convolutional neural networks. In: Advances in Neural Information Processing Systems, pp. 1097–1105 (2012)
8. McCloskey, M., Cohen, N.J.: Catastrophic interference in connectionist networks: the sequential learning problem. Psychol. Learn. Motiv. **24**, 109–165 (1989)
9. Radford, A., Metz, L., Chintala, S.: Unsupervised representation learning with deep convolutional generative adversarial networks. arXiv preprint arXiv:1511.06434 (2015)
10. Sermanet, P., Kavukcuoglu, K., Chintala, S., LeCun, Y.: Pedestrian detection with unsupervised multi-stage feature learning. In: Proceedings of the IEEE Conference on Computer Vision and Pattern Recognition, pp. 3626–3633 (2013)
11. Sutskever, I., Vinyals, O., Le, Q.V.: Sequence to sequence learning with neural networks. In: Advances in Neural Information Processing Systems, pp. 3104–3112 (2014)
12. Webb, G.I., Hyde, R., Cao, H., Nguyen, H.L., Petitjean, F.: Characterizing concept drift. Data Min. Knowl. Discov. **30**(4), 964–994 (2016)

Fault Prognostics for the Predictive Maintenance of Wind Turbines: State of the Art

Koceila Abid[1,2(✉)], Moamar Sayed Mouchaweh[1(✉)], and Laurence Cornez[2(✉)]

[1] IMT Lille Douai, Lille University, Lille, France
{koceila.abid,moamar.sayed-mouchaweh}@imt-lille-douai.fr
[2] CEA LIST, DM2I, LADIS, Gif-sur-Yvette Cedex, France
{koceila.abid,laurence.cornez}@cea.fr

Abstract. Reliability and availability of wind turbines are crucial due to several reasons. On the one hand, the number and size of wind turbines are growing exponentially. On the other hand, installation of these farms at remote locations, such as offshore sites where the environment conditions are favorable, makes maintenance a more tedious task. For this purpose, predictive maintenance is a very attractive strategy in order to reduce unscheduled downtime and maintenance cost. Prognostic is an online technique that can provide valuable information for proactive actions such as the current health state and the Remaining Useful Life (RUL). Several fault prognostic works have been published in the literature. This paper provides an overview of the different prognostic phases, including: health indicator construction, degradation detection, and RUL estimation. Different prognostic approaches are presented and compared according to their requirements and performance. Finally, this paper discusses the suitable prognostic approaches for the proactive maintenance of wind turbines, allowing to address the latter challenges.

Keywords: Fault prognostics · Remaining useful life · Predictive maintenance · Wind turbines

1 Introduction

The production of electricity using wind energy has an increasing trend in the last decade especially in Europe. It is reported in [1] that the wind energy has become the second source of energy in Europe behind the gas. This trend entails a growing evolution in the number and size of wind turbines (WTs), which lead to increase the cost of Operation and Maintenance (O&M). The cost of O&M for one wind turbine is about 20–30% of overall lifetime costs of energy.

Traditional maintenance strategies such as curative maintenance or preventive maintenance have a main drawback that is the fault is undergone by the system. Prognostics and Health Management (PHM) or predictive maintenance provides an advanced maintenance strategy that can enhance the reliability and

© Springer Nature Switzerland AG 2019
A. Monreale et al. (Eds.): ECML PKDD 2018 Workshops, CCIS 967, pp. 113–125, 2019.
https://doi.org/10.1007/978-3-030-14880-5_10

availability while reducing unscheduled fault and maintenance cost of systems such as WT. Health monitoring of WTs is achieved by monitoring the performance of the WT (e.g., power curve) [10] or condition monitoring of components, in particular critical components as rotating machinery (e.g., generator). Fault prognostic is an important step of the predictive maintenance. It can be defined as the prediction of when a failure might take place or the estimation of the Remaining Useful Life (RUL), which is the time between degradation detection and failure threshold. It is an online technique that can provide valuable information for proactive actions allowing mitigating the fault consequences and/or scheduling maintenance steps. RUL prediction of critical components is a very challenging task due to sensor noise, complexity of the system, and prediction uncertainty caused by the switch between operating conditions, and environment variability (e.g., wind speed and direction).

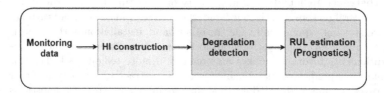

Fig. 1. General approach of prognostic.

The general approach of prognostic is presented in Fig. 1. It indicate the three main steps of prognostics: Health Indicator (HI) construction, Degradation detection and RUL estimation. First the HI is constructed by processing monitoring data in order to monitor the evolution of system performance over time. The second step is the degradation detection triggered as soon as the HI goes below a predefined threshold. The third step is the aims at predicting the degradation evolution and estimating the time when the system will go below a failure threshold. In this review, the techniques used to perform the different steps of the prognostic approach are presented and compared in terms of requirements and performances within the context of wind turbines predictive maintenance. The suitable prognostic approaches for the proactive maintenance of wind turbines are discussed, allowing to address the latter challenges.

2   Health Indicator Construction

Health Indicator (HI) construction is the main step for achieving prognostic. It represents the evolution over time of the system performance. When this evolution is decreasing, this indicates a drift from normal or nominal operation conditions towards a failure. Concept drift techniques can be used to monitor the performance of systems by learning patterns from data streams gathered by sensors [20], where a new class is detected if there is a decrease in the system performance according to its nominal value. HIs can be classified in two categories: physics health indicators and virtual health indicators which cannot be interpreted with physical sense. They can also be classified into HI based on a single feature such as using the raw data gathered from sensors, residuals based feature, time domain or time-frequency feature extracted from data measured by monitoring sensors. The second kind of HI is based on a fusion of multiple features that can perform better representation of the system health.

Single feature based HI is based on a single extracted feature which can be more interpretable. Most of prognostic works for wind turbine focused on the condition monitoring of WT rotating machine (e.g., low and high speed shaft, gearbox, generator) due to its criticity. Condition monitoring for such systems is achieved by using vibration signal, acoustic emission, oil analysis or current signal analysis. In the case of complex degradations, it is often hard to find one feature sensitive to those degradations. Therefore, it is unfeasible to construct a HI able to follow in the degradation evolution over time and to allow a reliable estimation of RUL. In order to tackle this problem, a solution is to fuse several features in order to exploit their complementarity. However, this fusion entails the lose of physics meaning (e.g., dimension reduction or distance between class in a feature space). This lose leads to a lack of HI interpretability so it represents a virtual description of the system performance health.

Several methods of HI construction based on a single and multiple features are proposed in the literature, which are summarized in Table 1.

2.1   Health Indicator Evaluation

HIs are evaluated using different criteria presented in [11]. The most pertinent HI evaluation criteria are monotonicity and trendability [9, 18, 36].

Monotonicity. The monotonicity evaluate the negative or positive trend of the HI, with the assumption that the system cannot self-heal [9, 18]. Monotonicity can be measured by the absolute difference between negative and positive derivative of HI as indicated in the following equation:

$$M = \left| \frac{\text{Number of } (d/dx > 0)}{n-1} - \frac{\text{Number of } (d/dx < 0)}{n-1} \right|, M \in [0; 1] \quad (1)$$

where d/dx represents the derivative of the HI, n represents the number of observations, M represents a higher monotonicity of a degradation when it approaches 1.

Trendability. Trendability is related to time and represents the correlation between the degradation trend and the operating time of a component [9], and can be calculated as follow:

$$R = \frac{n(\sum_{i=1}^{n} x_i t_i) - (\sum_{i=1}^{n} x_i)(\sum_{i=1}^{n} t_i)}{\sqrt{[n \sum_{i=1}^{n} x_i^2 - (\sum_{i=1}^{n} x_i)^2][n \sum_{i=1}^{n} t_i^2 - (\sum_{i=1}^{n} t_i)^2]}} \quad (2)$$

Table 1. Health indicator construction methods.

Feature type	Computation approaches	Methods	Data
Single feature based HI	Raw signal and residuals	Raw signal of viscosity and dielectric constant [37,38]	Lubrication oil (gearbox)
		Raw signal of oil debris [6]	Lubrication oil (gearbox)
		Power residual [28]	Generated power signal
		Temperature residual [2]	Bearings temperature (Gearbox)
	Time domain features	Root mean square [12,13]	Vibration (bearings)
		Spectral kurtosis [22]	Vibration (bearings)
		Trigonometric functions [9]	Vibration (bearings)
	Time-frequency features	Wavelet packet decomposition [24]	Vibration (bearings)
		Hilbert huang transform [18]	Vibration (bearings)
		Power spectral density [5]	Current signal (gearbox)
Multiple features based HI	Dimension reduction	ISOMAP [3,23]	Vibration (bearings)
		PCA [32]	SCADA data
	Distance between classes	Euclidian and mahalanobis distance [25,26]	Pitch angle (pitch system), voltage (converter)
		Jensen-Renyi distance [4]	Vibration (bearings)

$R \in [-1; 1]$ represents the correlation coefficient between indicator x and the time index t. R approaches 1, when the HI has a strong positive linear correlation with time.

3 Degradation Detection Based on Health Stage Division

The computed HI gives information about the system health (condition and performance). When a degradation occurred (drift from normal condition or nominal performance) the HI presents an increasing or decreasing trend. Then prognostic module can start because RUL prediction when no fault occurred is unnecessary. The degradation can be detected by dividing the HI into two or multiple stages using a threshold according to the degradation trend. Dividing the health stage using a threshold is widely used in the literature for this task, some works used classification techniques when sufficient degradation data are available in order to estimate the boundary between nominal and degraded conditions (see Table 2). Multiple stage division by considering an intermediate stage is also important in order to confirm the degradation occurrence with a second threshold for avoiding false alarms.

Table 2. Degradation detection methods.

Approaches	Methods	Advantage	Drawback
Threshold	Threshold value [15]	No need for degradation data. Easy for implementation	Difficulty to choose the threshold value in order to achieve early degradation detection and avoid false alarms. Need of an expert for the choice of the threshold value
	Statistical threshold 3σ interval [33]		
Classification	Logistic regression [34]	The boundary between nominal and degraded conditions is estimated automatically	Necessity of system degradation data. Model parameters are difficult to tune
	SVM [22]		

The goal of this division is to (1) identify the stages where the degradation process is active and (2) separate the degradation progress or evolution over time according to its dynamics (fast, slow, decreasing or self-healing, increasing and stable etc.). This division allows to improve the reliability of degradation detection and the RUL estimation.

4   Fault Prognostic (RUL Estimation)

Prognostic aims at estimating the remaining useful life of a component using the built HI. RUL estimation starts when a degradation is detected (drift from the nominal performance). Generally, RUL is estimated based on the use of one of the following two main approaches: Experience based approaches and degradation modeling approaches(see Fig. 2). Experience based prognostic is achieved by applying reliability or similarity based approaches. The degradation modeling can be achieved by using physical models or by data-driven approaches.

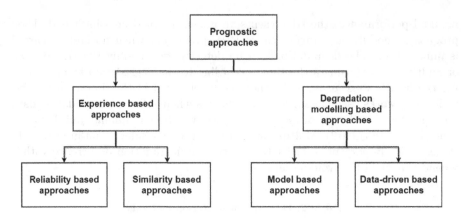

Fig. 2. RUL estimation approaches.

4.1 Experience Based Prognostic

Experience based prognostics methods are based on the use of maintenance or inspection feedback data sets gathered during a significant period of time. They represent the operational conditions during all the degradation process until the failure. In general, they can be classified into reliability based and similarity based approaches. Reliability based approaches employ a statistical failure distribution (i.e., Weibull) in order to fit to the gathered data. A periodic update of the parameters of this failure distribution is performed in order to adjust those parameters according to the experience obtained from the current maintenance practices. The built failure distribution is used for the RUL estimation of components with similar characteristics. Similarity based approaches use a library of degradation patterns for a set of components under different operation conditions. Each degradation pattern represents the degradation development of a component under certain operation conditions until the failure. The data gathered about the operation conditions of a component during a time window is compared to the degradation patterns in order to select the one who fits the best the data within this time window. Then, the RUL is estimated using the selected degradation pattern. Table 3 summarizes some experience based approaches presented in the literature for the prognostics.

Experience-based prognostics methods are efficient when degradation history is available for a system during a long period. The degradation history must represent the health evolution of the system from the healthy state until the failure state (run to failure). Reliability based approaches are not suitable for components showing complex degradation behaviour according to the operation and environment conditions. This is due to the fact that they do not use the online operation data in order to estimate the RUL. Similarity based approaches require also run to failure data history of identical units or components. Comparing to reliability approaches, similarity based approaches use online monitoring data in order to compare it with degradation history and estimate the RUL. They have

Table 3. Experience based approaches.

Approaches	Methods	Application	Advantage	Drawback
Reliability based approach	Weibull distribution [21,35]	Engine	Can monitor the system health state only with the operating cycle of the system	Depends on degradation history. Not adapted for complex systems due several interacting components and varying conditions. Less precise without using monitoring data
Similarity based approach	Similarity distance [29,30]	Electric cooling fan, engineered system	Good precision at component level if suitable data history is available. monitor the current health state of the system	Depends on degradation history. Not adapted for complex systems due several interacting components and varying conditions
	KNN [14,17]	Battery, turbofan, engineered system		
	Fuzzy similarity [39]	Reactor		

the advantage to be easy to implement and precise at the level of components if the latter present the same degradation trend. However, they are not adapted for complex systems due to the varying operation and environment conditions.

4.2 Degradation Modelling Based Approaches

This category of approaches aims at predicting the degradation evolution over time in order to estimate the system failure (end of life), Then the RUL is computed as the time from degradation detection until the failure. Model based approaches use physical and mathematical relations in order to model the degradation trend. These methods are usually used at component level and deal with wearing, crack, and corrosion phenomena. Paris law [16] is the most physics model used for the prediction of wear on rolling element bearing. Data driven approaches transform monitoring data gathered from sensors into relevant information about the system behaviour and dynamics, the mainly used methods are statistical and artificial Intelligence (AI) tools. Statistical techniques estimate the RUL of systems based on empirical knowledge, where artificial intelligence techniques are based on two phases: offline learning phase and online test phase. The model uses a database for training parameters in the offline phase, after that the online phase estimates the current health state of the system and predict its future state (evolution) over time. These methods are increasingly used due to their ability to find relation between data by training the model using datasets. Degradation modelling methods used in the literature are summarized in Table 4.

Table 4. Degradation modeling approaches.

Approaches		Methods	Application	Advantage	Drawback
Physical model based approach		Paris law [7, 8]	WT blades, gearbox bearings	Ability for interpretation by physics. Provide accurate RUL at component level prognostics	Detailed knowledge of the system behaviour is required. Complex systems degradation model is hard for construction
Data driven approach	Statistical methods	Moving average [6]	Gearbox	Understanding the degradation mechanism is not required. Effective in describing the uncertainty of the degradation process	Highly depends on the trend information of historical observations. Less precise in the RUL prediction of complex systems
		ARMA [34]	Engine		
		Bayesian filter [14]	Battery, turbofan engine		
		Particle filter [40]	Mechanical component		
	Artificial intelligence methods	RNN [9, 27]	Gearbox, Bearings	Able to learn complex nonlinear relationship between data. Understanding the degradation mechanism is not required. Expected to have a good performance in the RUL prediction of complex systems	Need sufficient data for training. Lack of physical meaning. Difficulty to select the parameters of the model
		ANFIS [31]	Gearbox shaft and gears		
		SVR [3, 18, 22]	Bearings		
		HMM [24]	Bearings		

Degradation modeling approaches use online monitoring data in order to observe the current health state of the system. Degradation evolution prediction is allowed when degradation of the system is detected. Model based degradation approaches are used generally at component level prognostic, where they gives more precision but they require extensive experimentation and model verification. These models are reliable when the system state have not been changed or upgraded. The models can be interpreted because its parameters are extracted from the system physics. However, it is difficult to generate degradation behaviour especially for complex system where several phenomena take place into the system. Data-driven based approaches are considered as black box models where the relation between input and output is complex and is hard for interpretation, but they can be used when the system is complex and where developing an accurate physical or mathematical model is not feasible. At component level, they can be less precise than the model based approaches, However, they are more suitable and efficient for complex systems whose degradation process is too hard to be modeled and represented by physical models.

5 Prognostic Metrics

When the RULs are predicted online for each sample of time, RULs must be evaluated using suitable and meaningful metrics. Root Mean Square Error (RMSE) and Mean Absolute Percentage Error (MAPE) are widely used in the literature for RUL evaluation depending on the true RUL.

$$RMSE = \sqrt{\frac{1}{n}\sum_{t=1}^{n}(r^l(t) - r_*^l(t))^2} \tag{3}$$

$$MAPE = \frac{100\%}{n}\sum_{t=1}^{n}\left|\frac{r^l(t) - r_*^l(t)}{r^l(t)}\right| \tag{4}$$

Where n is the number of observations, t is the time index, r_*^l represents the true RUL, and r^l represents the predicted RUL.

In [19], new metrics for prognostics performance evaluation are proposed. The most relevant metrics are: Prognostic Horizon, $\alpha - \lambda$ Performance, Relative Accuracy, and Cumulative Relative Accuracy. Prognostic Horizon (PH) is the difference between the current time index when the degradation is detected and the time of end of life (EoL). $\alpha - \lambda$ performance can determine whether the prediction falls within specified limits at particular times. Relative prediction accuracy is the errors between the predicted RUL relative to the actual (true) RUL at a specific time index. Cumulative Relative Accuracy CRA evaluates the RUL at multiple time instances. CRA is computed as the weighted sum of relative accuracies at multiple time instances.

6 Conclusion

The paper presents an overview of fault prognostic approaches for the monitoring and predictive maintenance of WTs. The fault prognostics of WTs is a challenging task because of their dynamics complexity, their different operating conditions, and the strong variability of their environment. The discussed prognostic approaches in this paper are compared according to their potential requirements in Table 5, where each requirement can be "Required", "Not required", and "Beneficial" for each prognostics approach.

Table 5. Prognostic approaches requirements.

Approaches	Engineering model	Degradation history	Current health state	Degradation detection
Reliability	Not required	Required	Not required	Not required
Similarity	Not required	Required	Required	Beneficial
Model based	Required	Beneficial	Required	Required
Data-driven	Not required	Beneficial	Required	Required

Table 6 presents a comparison between the four prognostics approaches in terms of precision, applicability, cost, and interpretability. For precision criteria, also component level and system level prognostic are compared for each prognostic approach, where (+) refers to the advantage and (−) refers to the drawback of the methods.

Table 6. Prognostic approaches comparison.

Approaches	Precision and applicability		Implementation	Cost	Interpretability
	Component level	System level			
Reliability	+	−	++	−−	−
Similarity	+	−	++	−−	−
Model	++	+	−	−	+
Data-driven	+	++	+	+	−

Experience based approaches (Reliability and similarity) are easier to apply when degradation history is available, but less precise at system level prognostic due to the variability of operation and environment conditions. Degradation modelling approaches are divided into model based and data driven approaches. Model based prognostic may have a good precision at component level (i.e., crack propagation of bearings). Although, when the system is more complex, this kind of approaches may not be precise or even applicable. Despite of the lack of interpretability of the data-driven approaches, they are the most suitable to perform the prognostic task of complex dynamic systems such as wind turbines.

Even with the increasing number of publications of fault prognostics for the predictive maintenance of WTs, several challenges still require to be addressed. Examples of these challenges are: how to ensure the prognostic when no a priori knowledge about the degradation behaviour is available (i.e., new installed system), which HI construction method is suitable for this case, the choose of the threshold value is also difficult and predicting the degradation evolution without a priori information about it is challenging.

The use of concept drift techniques can be an altenative solution to contribute addressing these challenges. They can be used in order to construct a HI and detect the drift from nominal conditions. When sufficient data about degradation is collected, the feature which fits the best the monotonicity and trendability during the degradation must be selected as HI. The threshold value must be chosen in order to detected the degradation as early as possible while avoiding the maximum of false alarms. In additions, multiple stage division should be considered in order to confirm the degradation occurrence. The degradation prediction model must also be updated with the collected incoming data in order to improve the performance of the prediction (precision). It is worth to mention that another challenge for fault prognostics is related to the RUL estimation for complex system due to the multiple interactions between their individual components. (see Fig. 3). Another challenge is related to the development of a post-prognostic phase in order to interpret prognostic results to the human operators by using a natural language.

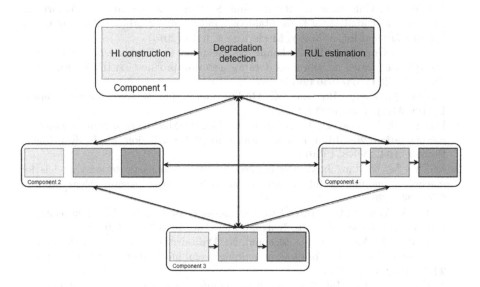

Fig. 3. Prognostic for complex system.

Acknowledgments. This work is supported by the European Union - European Regional Development Fund.

References

1. Annual combined onshore and offshore wind energy statistics. https://windeurope.org/about-wind/statistics/european/wind-in-power-2017. Accessed 26 June 2018
2. Bangalore, P., Tjernberg, L.B.: An artificial neural network approach for early fault detection of gearbox bearings. IEEE Trans. Smart Grid **6**(2), 980–987 (2015)
3. Benkedjouh, T., Medjaher, K., Zerhouni, N., Rechak, S.: Remaining useful life estimation based on nonlinear feature reduction and support vector regression. Eng. Appl. Artif. Intell. **26**(7), 1751–1760 (2013)
4. Boškoski, P., Gašperin, M., Petelin, D., Juričić, Ð.: Bearing fault prognostics using rényi entropy based features and Gaussian process models. Mech. Syst. Sig. Process. **52**, 327–337 (2015)
5. Cheng, F., Qu, L., Qiao, W.: Fault prognosis and remaining useful life prediction of wind turbine gearboxes using current signal analysis. IEEE Trans. Sustain. Energy **9**(1), 157–167 (2018)
6. Dupuis, R.: Application of oil debris monitoring for wind turbine gearbox prognostics and health management. In: Annual Conference of the Prognostics and Health Management Society, pp. 10–16 (2010)

7. Florian, M., Sørensen, J.D.: Wind turbine blade life-time assessment model for preventive planning of operation and maintenance. J. Mar. Sci. Eng. **3**(3), 1027–1040 (2015)
8. Grujicic, M., Galgalikar, R., Ramaswami, S., Snipes, J., Chenna, V., Yavari, R.: Finite-element analysis of horizontal-axis wind-turbine gearbox failure via tooth-bending fatigue. Int. J. Mater. Mech. Eng **3**, 6–15 (2014)
9. Javed, K., Gouriveau, R., Zerhouni, N., Nectoux, P.: Enabling health monitoring approach based on vibration data for accurate prognostics. IEEE Trans. Ind. Electron. **62**(1), 647–656 (2015)
10. Kusiak, A., Zheng, H., Song, Z.: On-line monitoring of power curves. Renew. Energy **34**(6), 1487–1493 (2009)
11. Lei, Y., Li, N., Guo, L., Li, N., Yan, T., Lin, J.: Machinery health prognostics: a systematic review from data acquisition to RUL prediction. Mech. Syst. Signal Process. **104**, 799–834 (2018)
12. Lei, Y., Li, N., Lin, J.: A new method based on stochastic process models for machine remaining useful life prediction. IEEE Trans. Instrum. Meas. **65**(12), 2671–2684 (2016)
13. Malhi, A., Yan, R., Gao, R.X.: Prognosis of defect propagation based on recurrent neural networks. IEEE Trans. Instrum. Meas. **60**(3), 703–711 (2011)
14. Mosallam, A., Medjaher, K., Zerhouni, N.: Data-driven prognostic method based on bayesian approaches for direct remaining useful life prediction. J. Intell. Manuf. **27**(5), 1037–1048 (2016)
15. Niu, G., Yang, B.S.: Intelligent condition monitoring and prognostics system based on data-fusion strategy. Expert. Syst. Appl. **37**(12), 8831–8840 (2010)
16. Paris, P., Erdogan, F.: A critical analysis of crack propagation laws. J. Basic Eng. **85**(4), 528–533 (1963)
17. Ramasso, E., Rombaut, M., Zerhouni, N.: Joint prediction of continuous and discrete states in time-series based on belief functions. IEEE Trans. Cybern. **43**(1), 37–50 (2013)
18. Saidi, L., Ali, J.B., Bechhoefer, E., Benbouzid, M.: Wind turbine high-speed shaft bearings health prognosis through a spectral kurtosis-derived indices and SVR. Appl. Acoust. **120**, 1–8 (2017)
19. Saxena, A., et al.: Metrics for evaluating performance of prognostic techniques. In: International Conference on Prognostics and Health Management, PHM 2008, pp. 1–17. IEEE (2008)
20. Sayed-Mouchaweh, M.: Learning from Data Streams in Dynamic Environments. Springer, Heidelberg (2016). https://doi.org/10.1007/978-3-319-25667-2
21. Schömig, A.K., Rose, O.: On the suitability of the Weibull distribution for the approximation of machine failures. In: Proceedings of IIE Annual Conference, p. 1. Institute of Industrial and Systems Engineers (IISE) (2003)
22. Soualhi, A., Medjaher, K., Zerhouni, N.: Bearing health monitoring based on Hilbert-Huang transform, support vector machine, and regression. IEEE Trans. Instrum. Meas. **64**(1), 52–62 (2015)
23. Tenenbaum, J.B., De Silva, V., Langford, J.C.: A global geometric framework for nonlinear dimensionality reduction. Science **290**(5500), 2319–2323 (2000)
24. Tobon-Mejia, D.A., Medjaher, K., Zerhouni, N., Tripot, G.: A data-driven failure prognostics method based on mixture of Gaussians hidden Markov models. IEEE Trans. Reliab. **61**(2), 491–503 (2012)
25. Toubakh, H., Sayed-Mouchaweh, M.: Hybrid dynamic data-driven approach for drift-like fault detection in wind turbines. Evol. Syst. **6**(2), 115–129 (2015)

26. Toubakh, H., Sayed-Mouchaweh, M.: Hybrid dynamic classifier for drift-like fault diagnosis in a class of hybrid dynamic systems: application to wind turbine converters. Neurocomputing **171**, 1496–1516 (2016)

27. Tse, P., Atherton, D.: Prediction of machine deterioration using vibration based fault trends and recurrent neural networks. J. Vib. Acoust. **121**(3), 355–362 (1999)

28. Uluyol, O., Parthasarathy, G., Foslien, W., Kim, K.: Power curve analytic for wind turbine performance monitoring and prognostics. Annu. Conf. Progn. Health Manag. Soc. **2**, 1–8 (2011)

29. Wang, P., Youn, B.D., Hu, C.: A generic probabilistic framework for structural health prognostics and uncertainty management. Mech. Syst. Sig. Process. **28**, 622–637 (2012)

30. Wang, T., Yu, J., Siegel, D., Lee, J.: A similarity-based prognostics approach for remaining useful life estimation of engineered systems. In: International Conference on Prognostics and Health Management, PHM 2008, pp. 1–6. IEEE (2008)

31. Wang, W.Q., Golnaraghi, M.F., Ismail, F.: Prognosis of machine health condition using neuro-fuzzy systems. Mech. Syst. Signal Process. **18**(4), 813–831 (2004)

32. Wang, Y., Ma, X., Joyce, M.J.: Reducing sensor complexity for monitoring wind turbine performance using principal component analysis. Renew. Energy **97**, 444–456 (2016)

33. Wang, Y., Peng, Y., Zi, Y., Jin, X., Tsui, K.L.: A two-stage data-driven-based prognostic approach for bearing degradation problem. IEEE Trans. Ind. Inform. **12**(3), 924–932 (2016)

34. Yan, J., Koc, M., Lee, J.: A prognostic algorithm for machine performance assessment and its application. Prod. Plan. Control. **15**(8), 796–801 (2004)

35. Zhai, L.Y., Lu, W.F., Liu, Y., Li, X., Vachtsevanos, G.: Analysis of time-to-failure data with Weibull model in product life cycle management. In: Nee, A., Song, B., Ong, S.K. (eds.) Re-engineering Manufacturing for Sustainability, pp. 699–703. Springer, Heidelberg (2013). https://doi.org/10.1007/978-981-4451-48-2_114

36. Zhang, B., Zhang, L., Xu, J.: Degradation feature selection for remaining useful life prediction of rolling element bearings. Qual. Reliab. Eng. Int. **32**(2), 547–554 (2016)

37. Zhu, J., Yoon, J., He, D., Qiu, B., Bechhoefer, E.: Online condition monitoring and remaining useful life prediction of particle contaminated lubrication oil. In: 2013 IEEE Conference on Prognostics and Health Management (PHM), pp. 1–14. IEEE (2013)

38. Zhu, J., Yoon, J.M., He, D., Qu, Y., Bechhoefer, E.: Lubrication oil condition monitoring and remaining useful life prediction with particle filtering. Int. J. Progn. Health Manag. **4**, 124–138 (2013)

39. Zio, E., Di Maio, F.: A data-driven fuzzy approach for predicting the remaining useful life in dynamic failure scenarios of a nuclear system. Reliab. Eng. Syst. Saf. **95**(1), 49–57 (2010)

40. Zio, E., Peloni, G.: Particle filtering prognostic estimation of the remaining useful life of nonlinear components. Reliab. Eng. Syst. Saf. **96**(3), 403–409 (2011)

Author Index